GW00818491

QUANTUM THEORY OF MANY-VARIABLE SYSTEMS AND FIELDS

Forthcoming titles in this series

Vol 2 — Relativistic Nuclear Physics: Theories of Structure and Scattering
L S Celenza & C Shakin

Vol 3 — The Landau Theory of Phase Transitions: Applicability to Structural, Incommensurate, Magnetic and Liquid Crystal Systems
J C Tolédano & P Tolédano

World Scientific Lecture Notes in Physics Vol. 1

QUANTUM THEORY OF MANY-VARIABLE SYSTEMS AND FIELDS

B Sakita (CUNY)

World Scientific

Published by

World Scientific Publishing Co. Pte. Ltd.
P. O. Box 128, Farrer Road, Singapore 9128

QUANTUM THEORY OF MANY-VARIABLE SYSTEMS AND FIELDS

ISBN 9971-978-55-5
9971-978-57-1 (pbk)

Printed in Singapore by Singapore National Printers (Pte) Ltd.

PREFACE

The present lecture notes are based on several special topics courses on Field Theory and Statistical Mechanics given at the City College of New York. The notes are compiled by the students and some of the calculations left out in the class were completed by them. I acknowledge J. Alfaro, A. Guha, W. Gutierrez, E. Gozzi, D.-X. Li, J. Malinsky, P. K. Mohapatra, for their participation in this endeavor.

Most of the materials chosen in these lectures are taken from the working examples completed when we tried to develop the non-perturbative methods in field theories. I acknowledge the collaboration with J.-L. Gervais and A. Jevicki for most of the work. I also acknowledge the collaboration with K. Kikkawa on a book written in Japanese on the same subjects. I made some conscious efforts not to overlap the contents, but it is impossible to do so since the basic material in these books was taken from the same set of notes.

Originally I planned to include one more chapter, XI. WKB III. General Theory, which is based on the work done with de Vega and Gervais. Unfortunately, however, I simply did not have enough time to prepare it in a satisfactory form. Instead, I included the most recent lecture note on Stochastic Quantization as the last part of the Appendix.

In preparing the final form of the notes I had the privilege of using the computer program set up by M. Lax. I thank him for permission to use his program and his associates, especially C. L. Wang, for his technical assistance and in the preparation of index. I also thank R. Tzani for her critical reading of the entire manuscript and suggesting many corrections. Finally I thank Mrs. E. De Crescenzo for typing the manuscript.

CONTENTS

I. Canonical Operator Formalism of Quantum Mechanics

1.1 Canonical Quantization

Classical Mechanics:

Let $L(\dot{q}, q)$ be a Lagrangian of a system, q being a dynamical variable and $\dot{q}$ its time derivative. The canonical momentum p is defined by

$$p = \frac{\partial L}{\partial \dot{q}} \tag{1.1}$$

and Hamiltonian of the system is given by the following Lagrange transform:

$$H(p, q) = p\,\dot{q} - L(\dot{q}, q) \tag{1.2}$$

The Hamiltonian is a function of q and p only, because

$$\delta H = \delta(p\,\dot{q} - L) = (\delta p\,\dot{q} + p\,\delta\dot{q} - \frac{\partial L}{\partial q}\delta q - \frac{\partial L}{\partial \dot{q}}\delta\dot{q})$$

$$= (\delta p\,\dot{q} - \frac{\partial L}{\partial q}\delta q) \tag{1.3}$$

The Lagrange equation of motion

$$\frac{d}{dt}\left(\frac{\partial L}{\partial \dot{q}}\right) - \frac{\partial L}{\partial q} = 0 \tag{1.4}$$

is a consequence of Hamilton's principle:

$$\delta \int L\, dt = 0 \tag{1.5}$$

The Hamilton equations of motion are then derived from (1.1), (1.3) and (1.4) :

$$\dot{q} = \frac{\partial H}{\partial p}, \qquad \dot{p} = -\frac{\partial H}{\partial q} \tag{1.6}$$

The equation of motion for an arbitrary physical quantity F , is then obtained from (1.6) as

$$\dot{F} = \frac{\partial F}{\partial q}\dot{q} + \frac{\partial F}{\partial p}\dot{p} = \frac{\partial H}{\partial p}\frac{\partial F}{\partial q} - \frac{\partial F}{\partial p}\frac{\partial H}{\partial q} = [H, F]_P \tag{1.7}$$

where $[\ ,\ ]_P$ is the Poisson bracket defined by

$$[A, B]_P = -[B, A]_P = \frac{\partial A}{\partial p}\frac{\partial B}{\partial q} - \frac{\partial B}{\partial p}\frac{\partial A}{\partial q} \tag{1.8}$$

Quantum Mechanics :

In canonical operator formalism of quantum mechanics the dynamical variable $\hat{q}$ and its canonical conjugate momentum $\hat{p}$ are operators in a Hilbert space (from here on operators are denoted with ^) and satisfy the following canonical commutation relation:

$$[\hat{q}, \hat{p}] \equiv \hat{q}\hat{p} - \hat{p}\hat{q} = i \tag{1.9}$$

The state of the system is a time dependent vector in the Hilbert space (Schrödinger picture), and the mechanical equation of the state vector is the Schrödinger equation:

$$i\frac{\partial}{\partial t}\ |\psi(t)\rangle = H(\hat{p}, \hat{q})\ |\psi(t)\rangle \tag{1.10}$$

where H is a Hamiltonian operator obtained from the classical Hamiltonian (1.2) by promoting the classical variables to the quantum operator variables.

In this procedure there exists an ambiguity if p and q appear in a product form because of the noncommutativity of $\hat{p}$ and $\hat{q}$. If this is the case one must define the quantum mechanics by specifying the order of operators. Accordingly, to a classical system many quantum mechanical systems may correspond.

The operator ordering ambiguity may not be as serious a problem for systems of few degrees of freedoms. For systems of many degrees of freedom especially for field theories, however, this is a serious problem because the different ordering may produce different interaction vertices.

In this lecture, therefore, we assume that the Hamiltonian has the following standard form:

$$H = \frac{1}{2}\sum_{m=1}^{M} p_m^2 + V(q_1, q_2, \ldots\ldots, q_M) \tag{1.11}$$

which is of course free of operator ordering ambiguity and the quantization is unique.

1.2 Heisenberg Picture

In Schrödinger picture operators $\hat{q}$ and $\hat{p}$ are time independent while the state vector is time-dependent.

The Heisenberg picture is a picture in which the operators are time-dependent and the state vector is not. To be an equivalent quantum mechanical description the Heisenberg picture should be related to the Schrödinger picture by a time-dependent unitary transformation:

$$\hat{q}(t) = \hat{U}^{+}(t)\, \hat{q}\, \hat{U}(t), \qquad \hat{p}(t) = \hat{U}^{+}(t)\, \hat{p}\, \hat{U}(t) \tag{1.12}$$

$$|\, \Phi > = \hat{U}^{+}(t)\, |\, \psi(t)> \tag{1.13}$$

$$\hat{U}(t)\hat{U}^{+}(t) = \hat{U}^{+}(t)\hat{U}(t) = 1 \tag{1.14}$$

Using Schrödinger equation (1.10) we obtain the equation for $\hat{U}(t)$:

$$i\frac{\partial}{\partial t}\hat{U}(t) = \hat{H}\, \hat{U}(t) \tag{1.15}$$

A formal solution of $\hat{U}$ is

$$\hat{U}(t) = e^{-i\hat{H}t} \tag{1.16}$$

The coordinate representation is the representation in which $\hat{q}$ is diagonal:

$$\hat{q}\, |\, q > = |\, q > q \tag{1.17}$$

$$< q' \,|\, q > = \delta(q - q')$$

The Schrödinger wave function is a component of $|\, \psi(t)>$ in $|\, q >$ basis :

$$\psi(q, t) \equiv < q\, |\, \psi(t)> = < q\, |\, \hat{U}(t)\, |\, \Phi > \tag{1.18}$$

Similarly, in the Heisenberg picture we consider a moving basis $|\, q, t >$ such that the Schrödinger wave function $\psi(q, t)$ is a component of $|\, \Phi >$ in $|\, q, t >$ basis:

$$\psi(q,t) = \langle q,t \mid \Phi \rangle \tag{1.19}$$

Comparing with (1.18) we obtain

$$\mid q,t \rangle = \hat{U}^{+}(t) \mid q \rangle \tag{1.20}$$

The interpretation of $\psi(q,t)$ we adapt is the standard probability amplitude interpretation. The transition probability $q,t \rightarrow q',t'$ is then given by

$$\langle q',t' \mid q,t \rangle = \langle q' \mid \hat{U}(t')\,\hat{U}^{+}(t) \mid q \rangle = \langle q' \mid e^{-i(t'-t)\hat{H}} \mid q \rangle \tag{1.21}$$

Thus, we call $\hat{U}(t)$ the evolution operator.

1.3 Interaction Picture

We first split the Hamiltonian into two parts, free and interaction:

$$\hat{H} = \hat{H}_0 + \hat{H}_1 \tag{1.22}$$

We call $\hat{H}_0$ "free" part. But it need not be a free Hamiltonian provided it is a Hamiltonian for which we can solve the problem exactly. Next, construct the evolution operator due to H_0:

$$\hat{U}_0(t) = e^{-i\hat{H}_0 t} \tag{1.23}$$

The interaction representation is defined by

$$\hat{q}_I(t) = \hat{U}_0^{+}(t)\,\hat{q}\,\hat{U}_0(t), \qquad \hat{p}_I(t) = \hat{U}_0^{+}(t)\,\hat{p}\,\hat{U}_0(t) \tag{1.24}$$

$$\mid \psi_I(t) \rangle = \hat{U}_0^{+}(t) \mid \psi(t) \rangle \tag{1.25}$$

Using (1.23) and the Schrödinger equation we obtain

$$i\frac{\partial}{\partial t} \mid \psi_I(t) \rangle = \hat{H}_I(t) \mid \psi_I(t) \rangle \tag{1.26}$$

where

$$\hat{H}_I(t) = \hat{U}_0^{+}(t)\,\hat{H}_1\,\hat{U}_0(t) \tag{1.27}$$

A formal solution of (1.26) is given by

$$|\psi_I(t)\rangle = T\, e^{-i\int_{-\infty}^{t}\hat{H}_I(t')dt'}\,|\psi_I(-\infty)\rangle$$

$$\equiv \hat{U}_I(t,-\infty)\,|\psi_I(-\infty)\rangle \qquad (1.28)$$

where T is the time ordering symbol defined by

$$T(\hat{H}_I(t)\hat{H}_I(t')) = \hat{H}_I(t)\hat{H}_I(t') \quad \text{for } t > t'$$

$$= \hat{H}_I(t')\hat{H}_I(t) \quad \text{for } t < t' \qquad (1.29)$$

etc.

Let $|n\rangle$ be an eigenstate of H_0 with eigenvalue E_n :

$$\hat{H}_0\,|n\rangle = E_n\,|n\rangle \qquad (1.30)$$

$|\langle n|\psi_I(t)\rangle|^2$ is the probability of finding the system in n-state at t . Suppose at $t = -\infty$ the state is prepared in $|i\rangle$ (initial) state:

$$|i\rangle = |\psi_I(-\infty)\rangle \qquad (1.31)$$

The probability amplitude of finding the system in $|f\rangle$ (final) at $t = \infty$ is given by

$$\langle f|\psi_I(\infty)\rangle = \langle f|\hat{U}_I(\infty,-\infty)|i\rangle \equiv \langle f|\hat{S}|i\rangle \qquad (1.32)$$

$\hat{S}$ is called the scattering operator (S-operator). Using (1.28) we obtain

$$\hat{S} = T\, e^{-i\int_{-\infty}^{\infty}\hat{H}_I(t)dt} \qquad (1.33)$$

Using (1.21) and (1.25) we also obtain

$$\hat{S} = \lim_{\substack{t'\to\infty \\ t\to-\infty}} e^{i\hat{H}_0 t'}e^{-i\hat{H}(t'-t)}e^{-i\hat{H}_0 t} \qquad (1.34)$$

1.4 Quantum Theory of Fields

The extension of the formalism described in the previous sections into many variables is trivially done by attaching an appropriate index to q and p .

$$q^m, \quad p^m \qquad (m = 1, 2, \ldots\ldots, M) \qquad (1.35)$$

$$[\hat{q}^m, \hat{p}_n] = i\,\delta_{mn} \tag{1.36}$$

Field theories are systems of many (infinite) degrees of freedom. As an example let us consider a real scalar field theory whose Lagrangian density is given by

$$\mathcal{L} = \frac{1}{2}(\partial_\mu\phi\partial^\mu\phi - m^2\phi) \tag{1.37}$$

where ϕ is a real scalar field which is a function of space-time point.

We restrict the space to a large finite volume V and $\phi(\vec{x},t)$ to satisfy the periodic boundary condition. We then expand ϕ into Fourier components:

$$\phi(\vec{x},t) = \frac{1}{V^{1/2}}\sum_{\vec{k}} e^{i\vec{k}\cdot\vec{x}} q_{\vec{k}}(t) \tag{1.38}$$

where momentum $\vec{k}$ is given by a set of integers n_x, n_y and n_z :

$$\vec{k} = \frac{2\pi}{L}\vec{n} \qquad (L^3 = V) \tag{1.39}$$

and $\sum_{\vec{k}}$ is the sum over n's. Note

$$q^*_{\vec{k}}(t) = q_{-\vec{k}}(t) \tag{1.40}$$

The Lagrangian of the system is then given by

$$L = \int_V d\vec{x}\,\mathcal{L} = \frac{1}{2}\sum_{\vec{k}}[\,|\dot{q}_{\vec{k}}(t)|^2 - \omega_k^2\,|q_{\vec{k}}(t)|^2] \tag{1.41}$$

where

$$\omega_k^2 = \vec{k}^2 + m^2 \tag{1.42}$$

The Lagrangian (1.41) is equivalent to a system of free harmonic oscillators. Thus, the quantization is straightforward:

$$[\hat{q}_{\vec{k}}, \hat{p}_{\vec{k}'}] = i\,\delta_{\vec{k},-\vec{k}'} \tag{1.43}$$

where $p_{\vec{k}}$ is the canonical conjugate momentum :

$$p_{\vec{k}} = \dot{q}_{\vec{k}} \tag{1.44}$$

Using (1.38) and analogous expression for

$$\hat{\pi}(\vec{x}) = \frac{1}{V^{1/2}} \sum_{\vec{k}} e^{i\vec{k}\cdot\vec{x}} \hat{p}_{\vec{k}} \tag{1.45}$$

we obtain

$$[\hat{\phi}(\vec{x}), \hat{\pi}(\vec{x}')] = i\,\delta(\vec{x}-\vec{x}') \tag{1.46}$$

In deriving (1.46) we used

$$\frac{1}{V}\sum_{\vec{k}} = \frac{1}{(2\pi)^3}\int d\vec{k} \tag{1.47}$$

which is valid in the infinite volume limit.

The Hamiltonian of the system is given by

$$\hat{H} = \frac{1}{2}\sum_{\vec{k}} (\hat{p}_{\vec{k}}^{\,2} + \omega_k^2 \hat{q}_{\vec{k}}^{\,2}) \tag{1.48}$$

$$= \frac{1}{2}\int d\vec{x}\,(\hat{\pi}^2(\vec{x}) + (\vec{\nabla}\hat{\phi}(\vec{x}))^2 + m^2\hat{\phi}^2(\vec{x})) \tag{1.49}$$

$$\equiv \frac{1}{2}\int d\vec{x}\,\hat{\pi}^2(\vec{x}) + V[\hat{\phi}] \tag{1.50}$$

Although we started with the Lorentz invariant Lagrangian density (1.37) the canonical operator formalism is inherently non-covariant, since the time is treated in the canonical formalism entirely differently from the space. The shortcoming of the non-covariance is remedied to some extent in the Heisenberg picture:

$$\hat{\phi}(\vec{x},t) = e^{i\hat{H}t}\hat{\phi}(\vec{x})e^{-i\hat{H}t}, \qquad \hat{\pi}(\vec{x},t) = e^{i\hat{H}t}\hat{\pi}(\vec{x})e^{-i\hat{H}t}$$

Using the explicit form of $\hat{H}$ we obtain

$$\partial_t\hat{\phi}(\vec{x},t) = \hat{\pi}(\vec{x},t), \qquad \partial_t\hat{\pi}(\vec{x},t) = (\nabla^2 + m^2)\hat{\phi}(\vec{x},t)$$

accordingly a covariant equation follows:

$$(\partial_t^2 - \nabla^2 + m^2)\hat{\phi}(\vec{x},t) \equiv (\Box + m^2)\hat{\phi}(x) = 0 \tag{1.51}$$

We demonstrated the Lorentz covariance of Heisenberg picture for a free scalar field. Convince yourself this is so with interactions provided that the interaction Lagrangian does not involve space-time derivatives.

II. Path Integral Formalism

2.1 Path Integral in Quantum Mechanics

Representation of Feynman Kernel in Terms of Phase Space Path Integral:

The Feynman kernel is defined by a kernel which connects Schrödinger wave functions in two different times:

$$\psi(q,t) = \int dq' \, K(q,t;q',t') \, \psi(q',t') \tag{2.1}$$

It is the transition probability amplitude defined by (1.21):

$$K(q,t:q',t') = \langle q \mid e^{-i(t-t')\hat{H}} \mid q' \rangle$$

$$\equiv \langle q,t \mid q',t' \rangle \tag{2.2}$$

Divide the time interval $t - t'$ into N pieces:

$$t - t' = N\epsilon \tag{2.3}$$

Let

$$t_n = t' + n\epsilon \tag{2.4}$$

From the completeness relation

$$1 = \int dq \mid q \rangle \langle q \mid$$

we obtain

$$1 = \prod_{n=1}^{N-1} \int dq_n \mid q_n, t_n \rangle \langle q_n, t_n \mid \tag{2.5}$$

Inserting (2.5) into (2.2) and taking the limit $N \to \infty$ we obtain

$$K(q,t;q',t') = \lim_{N \to \infty} \int .. \int \left[\prod_{n=1}^{N-1} dq_n \right]$$

$$\langle q,t \mid q_{N-1}, t_{N-1} \rangle \langle q_{N-1}, t_{N-1} \mid \quad \rangle ....$$

$$\langle q_n, t_n \mid q_{n-1}, t_{n-1} \rangle .... \langle q_1, t_1 \mid q', t' \rangle \tag{2.6}$$

This is a multiple integral and the integration variables are specified by a set of infinite variables: $q_1, q_2, \ldots, q_{N-1}$ $(N \to \infty)$. Since to these variables corresponds a path shown in Fig. 2-1 , we call integral (2.6) a path integral

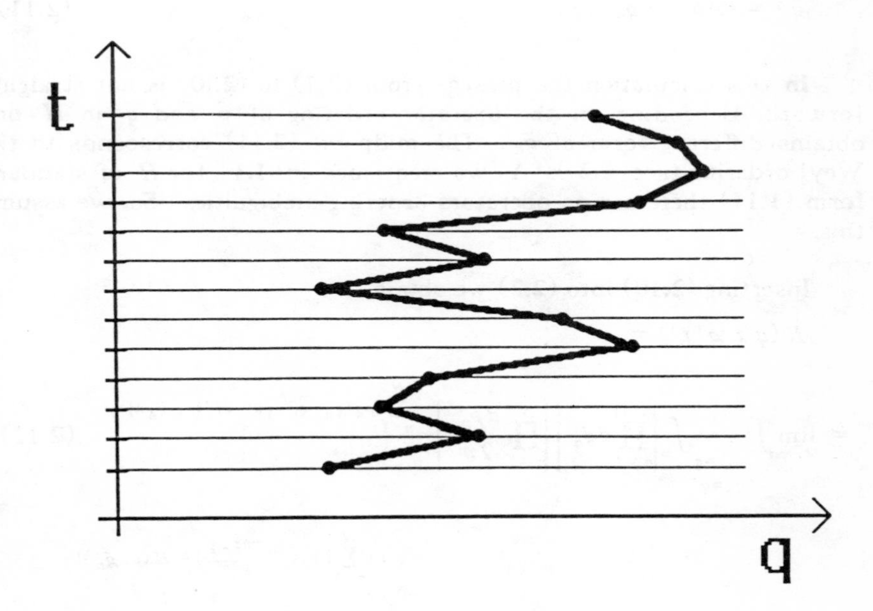

Fig. 2-1

Let us consider now a typical element in the integrand:

$$<q_n\, t_n \mid q_{n-1}\, t_{n-1}> = <q_n \mid e^{-i\epsilon \hat{H}(\hat{p}, \hat{q})} \mid q_{n-1}>$$

$$\approx <q_n \mid (1 - i\epsilon \hat{H}(\hat{p}, \hat{q})) \mid q_{n-1}> \qquad (2.7)$$

We then insert an identity

$$\int \frac{dp_n}{2\pi} \mid p_n><p_n \mid = 1 \qquad (2.8)$$

where $\mid p>$ is an eigenstate of $\hat{p}$ with eigenvalue p and

$$< q \mid p > = e^{ipq} \tag{2.9}$$

We obtain

$$< q_n, t_n \mid q_{n-1}, t_{n-1} > \approx \int \frac{dp_n}{2\pi} e^{ip_n(q_n - q_{n-1})} (1 - i\epsilon H(p_n, \bar{q}_n)) \tag{2.10}$$

where

$$\bar{q}_n = \tfrac{1}{2}(q_n + q_{n-1}) \tag{2.11}$$

In this calculation the passage from (2.7) to (2.10) is not straightforward. Depending on the operator ordering of $\hat{p}$ and $\hat{q}$ in H one obtains different form of $\bar{q}_n$. The midpoint (2.11) corresponds to the Weyl ordering (see **6.2**). As we mentioned in **1.1** , for H of standard form (1.11) there are no operators ordering ambiguities. So, we assume this.

Inserting (2.10) into (2.2) we obtain

$$K(q, t; q', t') =$$

$$= \lim_{N \to \infty} \int \cdots \int_{\substack{q_N = q \\ q_0 = q'}} \left[\prod_{n=1}^{N-1} dq_n \right] \left[\prod_{n=1}^{N} \frac{dp_n}{2\pi} \right] e^{i \sum_{n=1}^{N} [p_n(q_n - q_{n-1}) - \epsilon H(p_n, \bar{q}_n)]} \tag{2.12}$$

$$e^{i\epsilon \sum_{n-1}^{N} [p_n(\frac{q_n - q_{n-1}}{\epsilon}) - H(p_n, \bar{q}_n)]}$$

and we write K in a shorthand notation as

$$K(q, t; q', t') = \int \cdots \int_{\substack{q(t) = q, \\ q(t') = q'}} DqDp \; e^{i\int_{t'}^{t} dt [p(t)\dot{q}(t) - H(p(t), q(t))]}$$

$$= \int \cdots \int_{\substack{q(t) = q, \\ q(t') = q'}} DqDp \; e^{iA[p, q]} \tag{2.13}$$

where $A[p, q]$ is the action. This is the path integral in phase space, where q obeys the boundary condition:

$$q(t) = q, \qquad q(t') = q' \tag{2.14}$$

Because the number of p's and q's, over which we integrate are not the same the integration measure is not invariant by canonical transformations. The measur of path integral (2.13) deceptively looks as if it is invariant by canonical transformations. Special cares are needed to do the canonical transformations in path integral. (See **VI.**)

Feynman's Path Integral in Configuration (q-) Space:

Let us look again at the element

$$<q_n ,t_n \mid q_{n-1},t_{n-1}> = \int \frac{dp_n}{2\pi} e^{ip_n(q_n - q_{n-1}) - i\epsilon H(p_n,\bar{q}_n)}$$

$$= \int \frac{dp_n}{2\pi} e^{i\epsilon[p_n \frac{q_n - q_{n-1}}{\epsilon} - H(p_n,\bar{q}_n)]}$$

Assume that H has the standard form

$$H = \frac{1}{2}p^2 + V(q).$$

Then we can perform the integration over p_n :

$$\int \frac{dp_n}{2\pi} e^{i\epsilon[p_n \dot{q}_n - \frac{1}{2}p_n^2 - V(\bar{q}_n)]} = (2\pi i\epsilon)^{-1/2} e^{i\epsilon[\frac{1}{2}\dot{q}_n^2 - V(\bar{q}_n)]}$$

Inserting this into (2.6) we obtain

$$K(q,t;q',t') = \lim_{N\to\infty} \int .. \int \left| \prod_{n=1}^{N-1} dq_n \right| (2\pi i\epsilon)^{-N/2} e^{i\epsilon\sum_n(\frac{1}{2}\dot{q}_n^2 - V(\bar{q}_n))} \qquad (2.15)$$

In shorthand notation

$$<q \mid e^{-i\hat{H}(\hat{p},\hat{q})(t-t')} \mid q'> = \int_{\substack{q(t)=q \\ q(t')=q'}} \cdots \int Dq \, e^{i\int_{t'}^{t} dt L(q(t),\dot{q}(t))} \qquad (2.16)$$

This relation is the fundamental relation between the operator formalism and path integral representation. We obtained it for the standard form of H but as we shall see in **VI** that equation (2.16) holds for the general cases provided $\hat{H}$ is Weyl ordered and the path integral is defined by the midpoint prescription.

Exercise: Show that

$$<q_f, t_f \mid T(O_1(\hat{q}(t_1))O_2(\hat{q}(t_2))....) \mid q_i, t_i>$$

$$= \int_{\substack{q(t_i)=q_i \\ q(t_f)=q_f}} .. \int Dq\,(O_1(q(t_1)).O_2(q(t_2))....)\, e^{i\int_{t_i}^{t_f} dt\, L(q(t),\dot{q}(t))} \qquad (2.17)$$

where

$$t_f > t_1, t_2, ...> t_i$$

2.2 Path Integral in Quantum Field Theory

We emphasized in **I** that field theories are nothing but many variable systems. Therefore, the path integral formalism developed in the previous section can be extended to field theories (at least scalar field theories) by extending the formalism first to many variables. A special consideration is needed for Fermi fields, which we shall discuss in the next chapter. The canonical formalism of gauge fields and its path integral formulation will be discussed in the Appendix.

Drawback of the canonical quantum field theory in Schrödinger picture is its apparent Lorentz non-covariance. We now show that in the path integral formalism the transition amplitudes are expressed in a Lorentz covariant form.

Let $\mid\phi>$ be an eigenstate of scalar field operator $\hat{\phi}(\vec{x})$:

$$\hat{\phi}(\vec{x}) \mid \phi> = \mid \phi(\vec{x})> \phi(\vec{x}) \qquad (2.18)$$

To a given function of $\phi(\vec{x})$ corresponds a point in the configuration space. The Schrödinger wave function $<\phi \mid \psi(t)>$ is therefore a functional of $\phi(\vec{x})$. Feynman kernel is then defined by

$$K[\phi, t ; \phi', t'] = <\phi \mid e^{-i(t-t')\hat{H}} \mid \phi'> \qquad (2.19)$$

where $\hat{H}$ is the Hamiltonian operator and we assume it has a standard form:

$$\hat{H} = \frac{1}{2}\int \hat{\pi}^2(\vec{x})d\vec{x} + V[\hat{\phi}] \qquad (2.20)$$

For the case of free field $V[\phi] = \frac{1}{2}\int d\vec{x}((\nabla\hat{\phi})^2 + m^2\hat{\phi}^2)$ (see (1.50)).

The Feynman kernel (2.19) is then expressed in a path integral form as

$$<\phi|\, e^{-i(t-t')\hat{H}}\, |\phi'>$$

$$= \int_{\substack{\phi(\vec{x},t)=\phi(\vec{x}) \\ \phi(\vec{x},t')=\phi'(\vec{x})}} \cdots \int D\phi\, e^{i\int_{t'}^{t} dt\, (\frac{1}{2}\int \dot{\phi}^2(\vec{x},t)d\vec{x} - V[\phi])} \qquad (2.21)$$

The exponent of the path integrand is the action which is Lorentz invariant.

$$\frac{1}{2}\int \dot{\phi}^2 d\vec{x} - V[\phi] = \int d\vec{x}\frac{1}{2}[(\partial_\mu\phi)^2 + m^2\phi^2]$$

III. Path Integral Formalism of Fermi Fields

3.1 Grassmann Variables (anti-commuting c number variables)

We use ψ's and χ's for Grassmann variables, which have the following anti-commuting property:

$$\psi_i \psi_j = -\psi_j \psi_i \tag{3.1}$$

especially

$$\psi_i^2 = 0 \tag{3.2}$$

Let us restrict to a Grassmann variable ψ. We then define a function of ψ by

$$f(\psi) = f_0 + \psi f_1 \tag{3.3}$$

where f_0 and f_1 are ordinary numbers.

We define the derivative of ψ and the integral over ψ by

$$\frac{d}{d\psi} f(\psi) = f_1 \tag{3.4}$$

$$\int d\psi = 0, \qquad \int \psi d\psi = 1 \tag{3.5}$$

Note

$$\frac{d}{d\psi}[f(\psi)g(\psi)] = \left(\frac{d}{d\psi} f(\psi)\right) g(\psi) + f(-\psi)\left(\frac{d}{d\psi} g(\psi)\right) \tag{3.6}$$

namely

$$\frac{d}{d\psi}\psi = 1 - \psi\frac{d}{d\psi} \tag{3.7}$$

We also note

$$\int f(\psi) d\psi = f_1 = \left(\frac{d}{d\psi} f(\psi)\right) \tag{3.8}$$

so that

$$\int (\quad) d\psi = \frac{d}{d\psi} = \left(\frac{d}{d\psi}\right)^{-1} \tag{3.9}$$

The definition of integration (3.5) has similar properties of ordinary integration, namely

$$\int f(\psi+\chi)d\psi = \int f(\psi)d\psi \tag{3.10}$$

which corresponds to

$$\int_{-\infty}^{\infty} f(x+y)dx = \int_{-\infty}^{\infty} f(x)dx \tag{3.11}$$

and

$$\int \left(\frac{d}{d\psi} f(\psi) \right) d\psi = 0 \tag{3.12}$$

which corresponds to

$$\int_{-\infty}^{\infty} \left(\frac{d}{dx} f(x) \right) dx = f(\infty) - f(-\infty) = 0 \tag{3.13}$$

for $f(\pm\infty) = 0$.

For an ordinary complex number $z = x + iy$ one associates an independent complex number $\bar{z} = x - iy$ and calls it complex conjugate. We regard ψ as a complex Grassmann variable and associate to it an independent Grassmann variable $\bar{\psi}$, which we call conjugate:

$$\psi\bar{\psi} = -\bar{\psi}\psi$$

$$\frac{d}{d\psi}\bar{\psi} = -\bar{\psi}\frac{d}{d\psi} \tag{3.14}$$

etc.

A function of ψ and $\bar{\psi}$ and its integration are analogously defined:

$$F(\bar{\psi},\psi) = F_0 + \psi F_1 + \bar{\psi}\tilde{F}_1 + \psi\bar{\psi}F_2 \tag{3.15}$$

$$\int F(\bar{\psi},\psi)d\bar{\psi}d\psi = F_2 \tag{3.16}$$

where F_i 's are ordinary numbers.

Next let us consider a change of variables:

$$\psi \rightarrow \psi' = a\psi \ , \quad \bar{\psi} \rightarrow \bar{\psi}' = \bar{a}\bar{\psi} \tag{3.17}$$

where a is an ordinary number and $\bar{a}$ its complex conjugate. Integral (3.16) is then

$$\int F(\bar{\psi},\psi)d\,\bar{\psi}d\,\psi = \int F\left(\frac{\bar{\psi}'}{\bar{a}},\frac{\psi'}{a}\right)d\,\bar{\psi}'d\,\psi' J \tag{3.18}$$

where J is the Jacobian of transformation. The left hand side of (3.18) is F_2 while the right hand side is $\frac{F_2}{|a|^2}J$. So we obtain

$$J = |a|^2$$

The naive calculation of Jacobian, on the other hand, is given by

$$\frac{\partial(\bar{\psi},\psi)}{\partial(\bar{\psi}',\psi')} = \begin{vmatrix} \frac{\partial\bar{\psi}}{\partial\bar{\psi}'} & \frac{\partial\bar{\psi}}{\partial\psi'} \\ \frac{\partial\psi}{\partial\bar{\psi}'} & \frac{\partial\psi}{\partial\psi'} \end{vmatrix} = \begin{vmatrix} \frac{1}{\bar{a}} & 0 \\ 0 & \frac{1}{a} \end{vmatrix} = \frac{1}{|a|^2}$$

Therefore, we conclude that the Jacobian of Grassmann integration variables is inverse of the ordinary Jacobian:

$$\textit{Jacobian of Grassmann variables} = (\textit{ordinary Jacobian})^{-1} \tag{3.19}$$

We showed (3.19) for a simple transformation (3.17), but it can be proven in general.

The combination $\bar{\psi}\psi$ commutes with every number. We show next it can be treated as an ordinary number in the integration, namely

$$\int_{-\infty}^{\infty} f(x + \bar{\psi}\psi)dx = \int_{-\infty}^{\infty} f(x)dx \tag{3.20}$$

where we assume

$$f(\pm\infty) = 0 \tag{3.21}$$

and the integral is finite. The proof of (3.20) is straightforward:

$$\int_{-\infty}^{\infty}[f(x) + f'(x)\bar{\psi}\psi]dx = \int_{-\infty}^{\infty} f(x)dx + \bar{\psi}\psi[f(\infty) - f(-\infty)]$$

$$= \int_{-\infty}^{\infty} f(x)dx$$

We may extend ψ and $\bar{\psi}$ to a set of many variables, and consider a multiple integral:

$$\int \exp[-A(\bar{\psi}_1,\bar{\psi}_2,\ldots\ldots,\bar{\psi}_N,\psi_1,\psi_2,\ldots,\psi_N)]\prod_{i=1}^{N}(d\bar{\psi}_i d\psi_i) \qquad (3.22)$$

When we make a change of variables, we must insert Jacobian which is the inverse of the ordinary Jacobian. The integral is especially interesting and important when H has a quadratic form:

$$A[\bar{\psi},\psi] = \bar{\psi}_i \Omega_{ij} \psi_j \qquad (3.23)$$

Let us evaluate the integral for $N = 2$.

$$\int \exp[-(\bar{\psi}_1\Omega_{11}\psi_1 + \bar{\psi}_1\Omega_{12}\psi_2 + \bar{\psi}_2\Omega_{21}\psi_1 + \bar{\psi}_2\Omega_{22}\psi_2)]d\bar{\psi}_1 d\psi_1 d\bar{\psi}_2 d\psi_2$$

$$= \int \frac{1}{2}[\bar{\psi}_1\Omega_{11}\psi_1+\bar{\psi}_1\Omega_{12}\psi_2+\bar{\psi}_2\Omega_{21}\psi_1+\bar{\psi}_2\Omega_{22}\psi_2]^2\ldots\ldots$$

$$= \int [(\bar{\psi}_1\Omega_{11}\psi_1)(\bar{\psi}_2\Omega_{22}\psi_2) + (\bar{\psi}_1\Omega_{12}\psi_2)(\bar{\psi}_2\Omega_{21}\psi_1)]\ldots\ldots$$

$$= (\Omega_{11}\Omega_{22} - \Omega_{12}\Omega_{21})\int \bar{\psi}_1\psi_1\bar{\psi}_2\psi_2 d\bar{\psi}_1 d\psi_1 d\bar{\psi}_2 d\psi_2$$

$$= \det \Omega$$

From this calculation it is quite obvious that we get in general:

$$\int e^{-\bar{\psi}_i\Omega_{ij}\psi_j}\prod_{i=1}^{N} d\bar{\psi}_i d\psi_i = \det\Omega \qquad (3.24)$$

This result should be compared with the ordinary Gaussian integral:

$$\int \cdots \int e^{\bar{z}_i\Omega_{ij}z_j}\prod_{i=1}^{N}\frac{d\bar{z}_i dz_i}{\pi} = (\det\Omega)^{-1} \qquad (3.25)$$

3.2 Coherent State Representation of Fermi Operators in Terms of Grassmann Numbers

Review of Coherent State Representation of Bose Operators:

The coherent state representation of Bose operators is useful and it has been used not only in optics but also in quantum field theory (especially infrared problem in QED). We briefly review it.

Let $\hat{a}^+$ and $\hat{a}$ be creation and annihilation operators:

$$[\hat{a}, \hat{a}^+] = 1 \tag{3.26}$$

The coherent state representation is defined by

$$| z > = e^{z\hat{a}^+} | 0 > \tag{3.27}$$

which is an eigenstate of $\hat{a}$ with complex eigenvalue z.

$$\hat{a} \; | z > = z \; | z > \tag{3.28}$$

Note

$$\hat{a}^+ \; | z > = \frac{\partial}{\partial z} \; | z > \tag{3.29}$$

The adjoint of $| z >$ is defined by

$$< z | = < 0 | \, e^{\bar{z}\hat{a}} \tag{3.30}$$

The following expressions are useful:

$$< z | z' > = e^{\bar{z}z'} \tag{3.31}$$

$$< z | :h(\hat{a}^+, \hat{a}): | z' > = h(\bar{z}, z') e^{\bar{z}z'} \tag{3.32}$$

$$\int e^{-\bar{z}z} \, | z >< z | \, \frac{d\bar{z}dz}{2\pi} = 1 \tag{3.33}$$

where : : is the normal product symbol.

Coherent State Representation of Fermi Operators:

Let $\hat{\psi}^+$ and $\hat{\psi}$ be Fermi creation and annihilation operators:

$$[\hat{\psi}, \hat{\psi}^+]_+ \equiv \hat{\psi}\hat{\psi}^+ + \hat{\psi}^+\hat{\psi} = 1 \tag{3.34}$$

$$\hat{\psi}^2 = \hat{\psi}^{+2} = 0$$

The standard number representation is given by $| 0 >$ and $| 1 >$:

$$\hat{\psi}^+\hat{\psi} \; | 0 > = 0, \qquad \hat{\psi}^+\hat{\psi} \; | 1 > = | 1 > \tag{3.36}$$

$$\hat{\psi}|0\rangle = 0, \quad \hat{\psi}^{+}|0\rangle = |1\rangle, \quad \hat{\psi}|1\rangle = |0\rangle, \quad \hat{\psi}^{+}|1\rangle = 0 \tag{3.37}$$

The completeness relation is given by

$$|0\rangle\langle 0| + |1\rangle\langle 1| = 1 \tag{3.38}$$

By using a Grassmann number ψ we define a coherent state for Fermi operators by

$$|\psi\rangle \equiv |0\rangle + \psi|1\rangle = |0\rangle + \psi\hat{\psi}^{+}|0\rangle$$

$$= (1 + \psi\hat{\psi}^{+})|0\rangle = e^{\psi\hat{\psi}^{+}}|0\rangle \tag{3.39}$$

which has the following properties:

$$\hat{\psi}|\psi\rangle = \psi|\psi\rangle, \quad \hat{\psi}^{+}|\psi\rangle = \frac{\delta}{\delta\psi}|\psi\rangle \tag{3.40}$$

The adjoint coherent state is defined by

$$\langle\psi| \equiv \langle 0| + \langle 1|\bar{\psi} = \langle 0|e^{\hat{\psi}\bar{\psi}} \tag{3.41}$$

which has the properties

$$\langle\psi|\hat{\psi}^{+} = \langle\psi|\bar{\psi}, \quad \langle\psi|\hat{\psi} = \langle\psi|\frac{\delta}{\delta\bar{\psi}} \tag{3.42}$$

The analogous expressions to (3.31), (3.32) and (3.33) are

$$\langle\psi|\psi'\rangle = e^{\bar{\psi}\psi'} \tag{3.43}$$

$$\langle\psi|:h(\hat{\psi}^{+},\hat{\psi}):|\psi'\rangle = h(\bar{\psi},\psi')e^{\bar{\psi}\psi'} \tag{3.44}$$

$$\int |\psi\rangle\langle\psi|e^{-\bar{\psi}\psi}d\bar{\psi}d\psi = 1 \tag{3.45}$$

Exercise: Obtain (3.45) using (3.38).

3.3 Holomorphic Path Integral Representation

As shown in the previous section, the coherent representation of Bose and Fermi operators are formally the same. Thus the same path integral method can be used for Fermi and Bose systems. Let us confine ourselves to a Fermi system. We assume the Hamiltonian is normal ordered.

$$\hat{H} = :H(\hat{\psi}^+, \hat{\psi}): \tag{3.46}$$

The Feynman kernel is defined as usual by

$$K(\bar{\psi}_f, t_f; \psi_i, t_i) = <\psi_f | e^{-i\hat{H}(t_f - t_i)} | \psi_i > \tag{3.47}$$

Consider K as a multiple of many small evolutions.

$$e^{-i\hat{H}(t_f - t_i)} = \lim_{N\to\infty} (1 - i\epsilon\hat{H})^N \tag{3.48}$$

Thus,

$$K(\bar{\psi}_f, t_f; \psi_i, t_i) = \int \cdots \int \prod_{m=1}^{N-1} d\bar{\psi}_m d\psi_m$$

$$<\psi_f | (1 - i\epsilon\hat{H}(\hat{\psi}^+, \hat{\psi})) | \psi_{N-1} > e^{-\bar{\psi}_{N-1}\psi_{N-1}}$$

$$<\psi_{N-1} | (1 - i\epsilon\hat{H}) | \psi_{N-2} > \cdots$$

and

$$<\psi_{m+1} | (1 - i\epsilon H(\hat{\psi}^+, \hat{\psi})) | \psi_m > = (1 - i\epsilon H(\bar{\psi}_{m+1}, \psi_m)) e^{\bar{\psi}_{m+1}\psi_m}$$

$$\approx e^{-i\epsilon H(\bar{\psi}_{m+1}, \psi_m) + \bar{\psi}_{m+1}\psi_m} \tag{3.49}$$

Therefore, we obtain

$$K(\bar{\psi}_f, t_f; \psi_i, t_i) =$$

$$= \int \cdots \int \prod_{m=1}^{N-1} d\bar{\psi}_m d\psi_m e^{\sum_{n=1}^{N} [-i\epsilon H(\bar{\psi}_m, \psi_{m-1}) + \bar{\psi}_m \psi_{m-1}] - \sum_n^{N-1} \bar{\psi}_n \psi_n}$$

$$= \int \cdots \int \prod_{m=1}^{N-1} d\bar{\psi}_m d\psi_m e^{\sum_{n=1}^{N} [-\epsilon\bar{\psi}_m \frac{\psi_m - \psi_{m-1}}{\epsilon} - i\epsilon H(\bar{\psi}_m, \psi_{m-1})]} e^{\bar{\psi}_f \psi_f}$$

where

$$\psi_0 = \psi_i, \qquad \psi_N = \psi_f$$

The exponent of the integral can be expressed as

$$i\int_{t_i}^{t_f} dt\, [\, i\bar{\psi}(t)\dot{\psi}(t) - H(\bar{\psi}(t), \psi(t))\,] \equiv i\int_{t_i}^{t_f} dt L \tag{3.50}$$

Therefore finally we obtain

$$K(\bar{\psi}_f, t_f; \psi_i, t_i) = \int_{\substack{\psi(t_i)=\psi_i \\ \bar{\psi}(t_f)=\bar{\psi}_f}} \cdots \int D\bar{\psi} D\psi e^{i\int_{t_i}^{t_f} dt L} \tag{3.51}$$

3.4 Dirac Field

The quantum Hamiltonian of free Dirac field is given by

$$\hat{H} = \int d\vec{x} \hat{\psi}^+(\vec{x})(-i\,\vec{\alpha}\cdot\vec{\nabla} + m\,\beta)\hat{\psi}(\vec{x}) \tag{3.52}$$

where

$$[\hat{\psi}_\alpha(\vec{x}), \hat{\psi}_\beta^+(\vec{x}')]_+ = \delta_{\alpha\beta}\delta(\vec{x}-\vec{x}') \tag{3.53}$$

Applying the technique of the previous sections to this case, we write down the following path integral expression:

$$\int d\,\bar{\psi} d\,\psi\; e^{iA[\bar{\psi},\psi]} \tag{3.54}$$

where $A[\bar{\psi},\psi]$ is the action:

$$A = \int dt \int d\vec{x}\; [\, i\bar{\psi}\dot{\psi} - \bar{\psi}(-i\,\vec{\alpha}\cdot\vec{\nabla} + m\,\beta)\psi\,] \tag{3.55}$$

We then change variables

$$\bar{\psi} \to \bar{\psi}\gamma^0, \qquad \psi \to \psi$$

to obtain a covariant form:

$$A = \int d^4x\; \bar{\psi}(\, i\,\gamma^\mu\partial_\mu - m\,)\psi \tag{3.56}$$

[Note $\gamma^0 = \beta$, $\alpha^i = \gamma^0\gamma^i$]

IV. Perturbation Theory and Feynman Graphs

4.1 Generating Functional

Definition:

The basic quantity we like to compute is the Feynman kernel defined by (2.2) and expressed in terms of path integral by (2.16). From this, one obtains the physical transition amplitude by multiplying the initial and final wave function and integrating over the coordinates. In terms of path integral one expresses this amplitude as

$$<F \mid e^{-i\hat{H}(t_f - t_i)} \mid I> =$$

$$= \int \cdots \int Dq\, \psi_F{}^*(q(t_f))\psi_I(q(t_i))\, e^{i\int_{t_i}^{t_f} L(\dot{q}(t), q(t))dt} \tag{4.1}$$

where ψ_I and ψ_F are the initial and final wave function

$$<q \mid I> = \psi_I(q), \qquad <q \mid F> = \psi_F(q). \tag{4.2}$$

As we shall discuss in the Appendix, in field theories the S-matrix element is obtained by the reduction formula which relates the transition matrix element and the vacuum expectation value of the time ordered Heisenberg field operators -- Green's function:

$$G(x_1, x_2, \ldots, x_n) \equiv <0 \mid T(\hat{\phi}(x_1)\hat{\phi}(x_2)\ldots\hat{\phi}(x_n)) \mid 0> \tag{4.3}$$

where we denoted a space-time point by x. (From now on we use a simplified notation: x for $\vec{x}, t$ and dx for d^4x etc.)

One defines a generating functional of Green's function by

$$Z[j] = <0 \mid T\exp(i\int \hat{\phi}(x)j(x)dx) \mid 0> \tag{4.4}$$

so that Green's function (4.3) is obtained by the functional derivatives:

$$G(x_1, x_2, \ldots, x_n) = (-i)^n \left. \frac{\delta^n Z[j]}{\delta j(x_1)\delta j(x_2) \ldots \delta j(x_n)} \right|_{j=0} \tag{4.5}$$

The path integral expression of (4.4) is obtained by using (2.17).

$$Z[j] = \int \cdots \int D\phi \psi_0^* [\phi(\cdot,\infty)]\, \psi_0[\phi(\cdot,-\infty)]$$

$$\exp i \int dx\, (L(\phi(x), \partial_\mu \phi(x)) + \phi(x) j(x)) \quad (4.6)$$

where ψ_0 is the vacuum wave functional defined by

$$\psi_0 [\phi(\cdot)] = < \phi \mid 0 > \quad (4.7)$$

In the path integral expressions (4.1) and (4.6) the wave functions are contained in the integrand. In **V**, we shall discuss the path integral expression for the generating functional of correlation functions in statistical mechanics. There again one obtains a similar expression (see (5.8)). In order to cover all these cases we consider the generating functional

$$Z[j] = \int \cdots \int Dq\ \rho(q(t_f), q(t_i)) e^{i \int_{t_i}^{t_f} dt\, [L(\dot{q}(t), q(t)) + j(t) q(t)]} \quad (4.8)$$

from which we obtain the field theory case by first restricting L and ρ to

$$L(\dot{q}, q) = L_0(\dot{q}, q) + L_1(q). \qquad L_0 = \frac{1}{2}(\dot{q}^2 - \omega^2 q^2) \quad (4.9)$$

$$\rho(q(t_f), q(t_i)) = \psi_0^* (q(t_f)) \psi_0(q(t_i))$$

$$= (\omega / \pi)^{1/2} e^{-\frac{\omega}{2}(q^2(t_f) + q^2(t_i))} \quad (4.10)$$

and then by extending (4.8) to many variables. For the case of statistical mechanics which will be discussed in **V**, we set

$$\rho(q', q) = \delta(q' - q) \quad (4.11)$$

and change the time to the negative imaginary time.

Calculation of $Z_0[j]$:

We start with (4.8) with L given by (4.9). The following expression is quite obvious:

$$Z[j] = e^{i \int_{t_i}^{t_f} dt L_1(\frac{1}{i} \frac{\delta}{\delta q(t)})} Z_0[j] \quad (4.12)$$

where $Z_0[j]$ is given by (4.8) with L_0 instead of L

Next we change variables $q(t) \to x(t)$:

$$q(t) = x(t) + x_0(t) \tag{4.13}$$

where $x_0(t)$ is a functional of $j(t)$. In this change of variables we choose x_0 in such that all the j-dependence of Z_0 is factored out. We substitute (4.13) into the exponent of the integral

$$\int_{t_i}^{t_f} [[\frac{1}{2}\dot{q}^2(t) - \frac{\omega^2}{2}q^2(t)] + j(t)q(t)]\,dt$$

$$= \int_{t_i}^{t_f} [(\frac{1}{2}\dot{x}_0^2 - \frac{\omega^2}{2}x_0^2(t)) + j(t)x_0(t)]dt$$

$$+ \int_{t_i}^{t_f} (\frac{1}{2}\dot{x}^2(t) - \frac{\omega^2}{2}x^2(t))dt$$

$$+ \int_{t_i}^{t_f} [\dot{x}\dot{x}_0 - \omega^2 x x_0 + jx]dt \tag{4.14}$$

The first term in the right hand side of (4.14) contains only x_0 and j so that it depends only on j. On the other hand the second term contains only x, the integration variable. The third term contains both $x(t)$ and $x_0(t)$:

$$\int_{t_i}^{t_f} [\dot{x}\dot{x}_0 - \omega^2 x x_0 + jx]dt =$$

$$= -\int_{t_i}^{t_f} [\ddot{x}_0 + \omega^2 x_0 - j]x(t)dt + x(t_f)\dot{x}_0(t_f) - x(t_i)\dot{x}_0(t_i) \tag{4.15}$$

We note that in this expression the first term depends on $x(t)$ for all t while the last two terms depend only on the boundary values. Therefore, in order to factor out all the j-dependence we choose x_0 to be a solution of

$$(\partial_t^2 + \omega^2)x_0(t) = j(t) \tag{4.16}$$

which makes the first term disappear. But the second and third term give an extra phase factor:

$$e^{i(x(t_f)x_0(t_f) - x(t_i)x_0(t_i))} \tag{4.17}$$

For the case of (4.11) (periodic boundary condition) we simply set

$$x_0(t_i) = x_0(t_f), \qquad \dot{x}_0(t_i) = \dot{x}_0(t_f) \tag{4.18}$$

to eliminate the phase factor.

Next let consider the case of (4.10). By the change of variables we obtain

$$\psi_0^*(q(t_f))\psi_0(q(t_i)) =$$

$$= \psi_0^*(x(t_f))\psi_0(x(t_i))e^{-\frac{\omega}{2}(x_0^2(t_i) + x_0(t_f))} e^{-\omega(x(t_f)x_0(t_f) + x(t_i)x_0(t_i))} \tag{4.19}$$

Thus, if we impose the following boundary condition the last exponential factor is canceled with (4.17) and all the dependence at the boundary factorized:

$$i\dot{x}_0(t_f) = \omega x_0(t_f), \qquad i\dot{x}_0(t_i) = -\omega x_0(t_i) \tag{4.20}$$

The first term in the right hand side of (4.14) is now given by

$$\int [\tfrac{1}{2}\dot{x}_0^2 \mp \frac{\omega^2}{2}x_0^2 + jx_0]dt = \tfrac{1}{2}x_0\dot{x}_0\Big|_{t_i}^{t_f} + \tfrac{1}{2}\int dt \;\; jx_0$$

and the surface term (first term) is canceled with the first exponential factor of (4.19). Thus, in either case we obtain

$$Z_0[j] = \exp\left(\frac{i}{2}\int_{t_i}^{t_f} j(t)x_0(t)dt\right) Z_0[0] \tag{4.21}$$

Equation (4.18) and (4.20) serve as boundary conditions for the second order differential equation of $x_0(t)$: (4.16). The solution is written in general as

$$x_0(t) = -i\int_{t_i}^{t_f} dt'\Delta(t,t')j(t') \tag{4.22}$$

where $\Delta(t,t')$ is a Green's function. Inserting (4.22) into (4.21) we obtain

$$Z_0[j] = \exp\left[\frac{1}{2}\int\int_{t_i}^{t_f} dt dt'\; j(t)\Delta(t,t')j(t')\right] Z_0[0] \qquad (4.23)$$

4.2 Feynman Propagator

We take the boundary condition (4.20) and set

$$t_i \to -\infty, \qquad t_f \to +\infty$$

The corresponding Δ is denoted by $\Delta_F(t-t';\omega)$, which satisfies

$$(\partial^2 + \omega^2)\Delta_F(t-t';\omega) = -i\delta(t-t') \qquad (4.24)$$

$$i\partial_t \Delta_F(t-t';\omega)\Big|_{t=\pm\infty} = \pm\omega\Delta_F(t-t';\omega)\big|_{t=\pm\infty} \qquad (4.25)$$

The solution of these equations is given by

$$\Delta_F(t-t';\omega) = \frac{1}{2\omega}[\theta(t-t')e^{-i\omega(t-t')} + \theta(t'-t)e^{i\omega(t-t')}]$$

$$= \frac{i}{2\pi}\int_{-\infty}^{\infty} d\omega' \frac{e^{-i\omega'(t-t')}}{\omega'^2 - \omega^2 + i\epsilon} \qquad (4.26)$$

Notice (4.26) is the same as the two-point Green's function of free harmonic oscillator:

$$\Delta_F(t-t';\omega) = \langle 0|T(\hat{q}(t)\hat{q}(t'))|0\rangle \qquad (4.27)$$

We can extend the formalism to many variable systems by adding indices. For a real scalar field we use the expansion (1.38) and similar expansion for j. For a free Lagrangian given by (1.41) and (1.42), the corresponding exponent of (4.22) is

$$\frac{1}{2}\int\int dt dt' \sum_{\vec{k}} j(\vec{k},t)\Delta_F(t-t';\omega_k)j(-\vec{k},t')$$

$$= \frac{1}{2}\int\int dt dt' \int\int d\vec{x} d\vec{x}'\, j(\vec{x},t)j(\vec{x}',t')\frac{1}{V}\sum_{\vec{k}} e^{i(\vec{x}-\vec{x}')\cdot\vec{k}}\Delta_F(t-t';\omega_k)$$

$$= \frac{1}{2}\int\int d^4x d^4x'\, j(x)j(x')\Delta_F(x-x') \qquad (4.28)$$

where

$$\Delta_F(x-x') = \frac{1}{V}\sum_{\vec{k}} e^{i(\vec{x}-\vec{x}')\cdot\vec{k}}\Delta_F(t-t';\omega_k)$$

$$= \frac{i}{(2\pi)^4}\int d^4k \frac{e^{-ik(x-x')}}{k^2-m^2+i\epsilon} \tag{4.29}$$

Thus, we obtain

$$Z_0[j] = \int \cdots \int D\phi e^{i\int dx(L_0+\phi j)}$$

$$= \exp\left[\frac{1}{2}\int\int dxdx' j(x)\Delta_F(x-x')j(x')\right] Z_0[0] \tag{4.30}$$

Exercise: Work out (4.26), (4.28) and (4.29).

4.3 Perturbation Expansion and Feynman Rules

In this section we show the perturbation expansion for neutral scalar field theory with ϕ^4 interaction:

$$L_1 = -\frac{\lambda}{4!}\phi^4 \tag{4.31}$$

where λ is a coupling constant. We seek a power series expansion of λ for generating functional $Z[j]$.

First, we write $Z[j]$ by using (4.12) and (4.22) as

$$Z[j] = e^{i\int dx L_1\left(\frac{1}{i}\frac{\delta}{\delta j(x)}\right)} e^{\frac{1}{2}\int\int dxdx' j(x)\Delta_F(x-x')j(x')} \tag{4.32}$$

Then we use the identity

$$F\left(\frac{\partial}{i\,\partial x}\right)G(x) = G\left(\frac{\partial}{i\,\partial y}\right)F(y)e^{ixy}\Big|_{y=0} \tag{4.33}$$

to obtain

$$Z[j] = e^{-\frac{1}{2}\int\int dxdx' \Delta_F(x-x')\frac{\delta^2}{\delta\phi(x)\delta\phi(x')}}$$

$$e^{i\int dx(L_1(\phi(x)) + j(x)\phi(x))}\Big|_{\phi(x)=0} \tag{4.34}$$

The last exponential is expanded as

$$e^{i\int dx\,(L_1(\phi(x)) + j(x)\phi(x))} =$$

$$=\sum_{n,p} \frac{i^n}{n!}\frac{i^p}{p!}\int dx_1 \cdots \int dx_n \int dy_1 \cdots \int dy_p\, j(x_1) \cdots j(x_n)$$

$$L_1(\phi(y_1)) \cdots L_1(\phi(y_p))\phi(x_1) \cdots \phi(x_n) \qquad (4.35)$$

Inserting this expansion into (4.34) and performing the functional derivative we obtain a power series expansion of generating functional.

Since the n-point Green's function is obtained by (4.5) and since the interaction Lagrangian is given by (4.31). The n, p term in (4.35) contributes to the λ^p term of n-point Green's function. In (4.34) we set $\phi(x) = 0$ after all the derivative are taken, so that in the expansion of exponentials only the terms which contain equal number of ϕ's and $\frac{\delta}{\delta\phi}$'s survive. Let us denote by $Z^{(n,p)}[j]$ the corresponding term in the expansion of $Z[j]$ and analyze a few simple cases.

n = 2, p = 0:

In the expansion of first exponential in (4.34), only

$$-\frac{1}{2}\int dx \int dx'\Delta_F(x-x')\frac{\delta^2}{\delta\phi(x)\delta\phi(x')}$$

term survive.

$$Z^{(2,0)} = -\frac{1}{2}\int dx \int dx'\Delta_F(x-x')\frac{\delta^2}{\delta\phi(x)\delta\phi(x')}$$

$$\frac{i^2}{2!}\int dx_1 \int dx_2 j(x_1) j(x_2)\phi(x_1)\phi(x_2)$$

Since

$$\frac{\delta^2}{\delta\phi(x)\delta\phi(x')}\phi(x_1)\phi(x_2) = \delta(x-x_1)\delta(x'-x_2) + \delta(x-x_2)\delta(x'-x_1)$$

we obtain

$$Z^{(2,0)} = \frac{1}{2}\int dx_1 \int dx_2 j(x_1) j(x_2)\Delta_F(x_1-x_2) \qquad (4.36)$$

For the bookkeeping purpose we denote (4.36) by the diagram of Fig. 4-1.

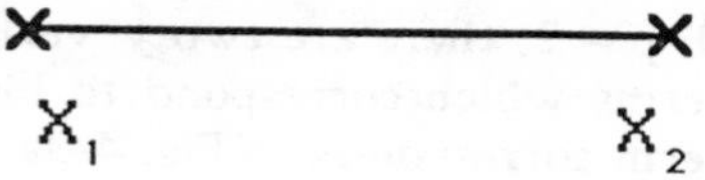

Fig. 4-1

X_1 X_2

n = 0 , p = 1:

The relevant term in the expansion (4.35) is

$$i\int dy_1 L_1(\phi(y_1)) = -i\frac{\lambda}{4!}\int dy_1\phi^4(y_1)$$

We need four $\frac{\delta}{\delta\phi}$'s so that

$$Z^{(0,1)} = \frac{1}{2!}\left(-\frac{1}{2}\int dx\int dx'\Delta_F(x-x')\frac{\delta^2}{\delta\phi(x)\delta\phi(x')}\right)^2\left(-i\frac{\lambda}{4!}\int dy_1\phi^4(y_1)\right)$$

There are 4! possible ways of making derivatives (contractions) which all give the same contributions. Thus,

$$Z^{(0,1)} = -i\frac{\lambda}{2\cdot 2\cdot 2}\int dy_1(\Delta_F(0))^2 \tag{4.37}$$

The corresponding diagram is shown in Fig. 4-2. We can read the combinatorial factor as 1/2 for two equivalent lines and 1/2 each for the lines which connect the same point.

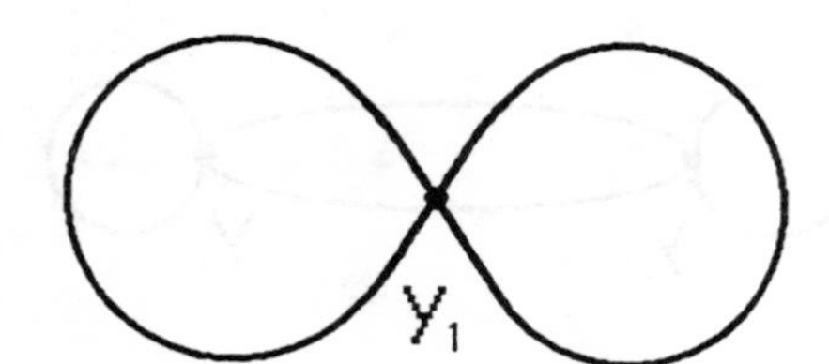

Fig. 4-2

n = 0 , p = 2:

By now it is clear that the bookkeeping diagram. Feynman diagram, consists of vertex points which corresponds to x's and y's in the expansion (4.35). For x's we attach $j(x)$ and denote it by a cross as shown in Fig. 4-1. One line is coming out of each x . On the other hand in $\lambda\phi^4$ theory four lines are coming out of each y-vertex as shown in Fig. 4-2. The derivative operations (contractions) are equivalent to connect these vertices by internal lines, to each of which we associate a Feynman propagator.

For the case of n = 0, p = 2, there are two y vertices and no x vertex. So, we expect two terms which correspond to Fig. 4-3a and Fig. 4-3b. Let us calculate the term corresponds to Fig. 4-3a explicitly.

$$Z^{(0,2)} = \frac{1}{2!} \int dy_1 \int dy_2 \frac{1}{4!} \left[\frac{1}{2} \int dx_1 \int dx' \Delta_F(x-x') \frac{\delta}{\delta\phi(x)} \frac{\delta}{\delta\phi(x')} \right]^4$$

$$\left[-i\frac{\lambda}{4!}\phi^4(y_1) \right] \left[-i\frac{\lambda}{4!}\phi^4(y_2) \right] \tag{4.38}$$

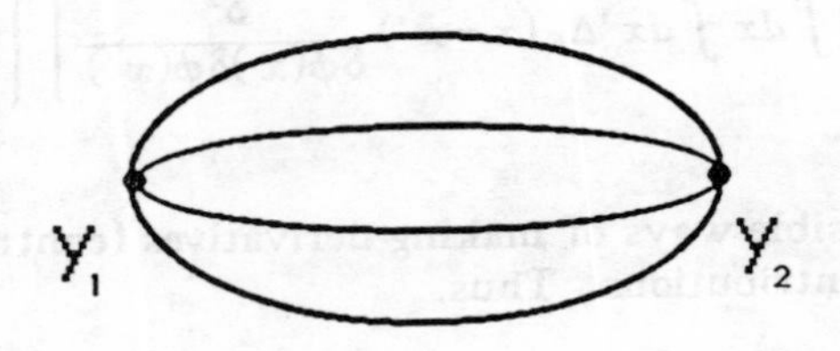

Fig. 4-3a

Fig. 4-3b

In order to connect vertices y_1 and y_2 vertex by four lines, we must operate $\frac{\delta}{\delta\phi(x)}$ and $\frac{\delta}{\delta\phi(x')}$ to different vertices. After a careful counting of combinations, we obtain

$$Z^{(0,2)}(Fig. 4\text{-}3a) = \frac{(-i\lambda)^2}{2!4!} \int dy_1 \int dy_2 \left[\Delta_F(y_1 - y_2) \right]^4 \tag{4.39}$$

One interprets the combinatorial factor as 1/2! for the symmetry of interchange of y_1 and y_2 (the same 1/2! appeared in Fig. 4-1.) and 1/4! for 4 equivalent lines between y_1 and y_2.

After this much work it is easy to generalize the argument to obtain the following standard Feynman rules:

1. Draw all distinct diagrams with n (even) x-vertices and p y-vertices, and sum all the contributions according to the following.
2. To each y-vertex attach a factor $-i\lambda$
3. To each x-vertex attach $ij(x)$
4. To each line between two vertex-points, say x and y, attach a propagator $\Delta_F(x-y)$
5. Multiply the combinatorial factor according to the following rule:
   (a) $1/m!$ if diagram is symmetric by the interchanges of m of x's or y's.
   (b) $1/m!$ for m equivalent internal lines
   (c) 1/2 for each closed line
   (d) $1/m!$ for m equivalent disconnected diagrams
6. Integrate x's and y's.

The rule 5(d) may require an explanation. The partition function for a diagram which contains several disconnected diagrams is given by a product of partition functions of connected diagrams. However, when it contains m identical connected diagrams one must multiply $1/m!$ (see **Exercise**).

Feynman Rules in Momentum Space:

Fourier transform of Green's function is defined by

$$\tilde{G}(p_1, p_2, \cdots p_n) = \int dx_1 dx_2 \cdots dx_n \, e^{-i\sum_{j=1}^{n} p_j \cdot x_j} G(x_1, x_2 \cdots x_n) \quad (4.40)$$

Note

$$Z[j] = \sum_n \frac{i^n}{n!} \int \cdots \int dx_1 \cdots dx_n \, j(x_1) \ldots j(x_n) \, G(x_1, \ldots, x_n) \quad (4.41)$$

which is the same statement as (4.5). Feynman rules for $G(x_1 \cdots x_n)$ are obtained accordingly. The Feynman rules for $\tilde{G}(p_1, \cdots, p_n)$ are obtained by making an appropriate Fourier transform. We use momenta p's for the momenta corresponding to the coordinates x's and k's for

y's.

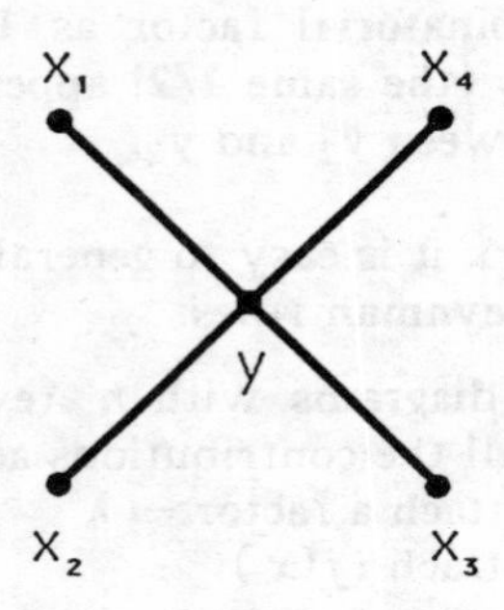

Fig. 4-4

Let us look at a y-vertex of Feynman graph Fig. 4-4, which we express as

$$-i\lambda \int dy\, \Delta_F(x_1-y)\Delta_F(x_2-y)\Delta_F(x_3-y)\Delta_F(x_4-y) \tag{4.42}$$

Using the integral representation of $\Delta_F(x)$, we perform y-integration to yield a δ-function, which represents a four momentum conservation at the vertex. Fourier transform of (4.42) is then given by

$$\prod_{i=1}^{4}\left(\frac{i}{p_i^2-m^2+i\epsilon}\right)\left(-i\lambda\delta^{(4)}(p_1+p_2+p_3+p_4)\right) \tag{4.43}$$

From this example it is quite obvious now to obtain the following Feynman rules in momentum space.

1. Draw all topologically distinct diagrams with n external lines. Each line carries a momentum. Denote them by p's for external lines and by k's for internal.
2. Assign the factor $\dfrac{i}{p^2-m^2+i\epsilon}$ for external line.
3. Assign the factor $\dfrac{d^4k}{(2\pi)^4}\dfrac{i}{k^2-m^2+i\epsilon}$ for an internal line.
4. To each vertex assign $-i\lambda(2\pi)^4\delta^{(4)}(q)$, where q is the sum of incoming momenta to the vertex.
5. Multiply the combinatorial factors, which are the same as coordinate space Feynman rules.
6 Integrate over k's.
7 Sum the contributions of all topologically distinct Feynman diagrams.

Exercise: Prove (4.33).

4.4 Proper Graphs and Theory of Effective Action

Connected Diagrams, Proper Diagrams:

The connected diagrams would need no explanation. By $W[j]$ we denote the generating functional of connected Green's functions. Then due to Feynman rule 5(d), we obtain

$$Z[j] = e^{W[j]} \tag{4.44}$$

All the connected diagrams, which can be made two disconnected diagrams by removing one internal line, are called improper (or one particle reducible). Non-improper connected diagrams are proper diagrams (or one particle irreducible). Fig. 4-5a is improper while Fig. 4-5b is proper.

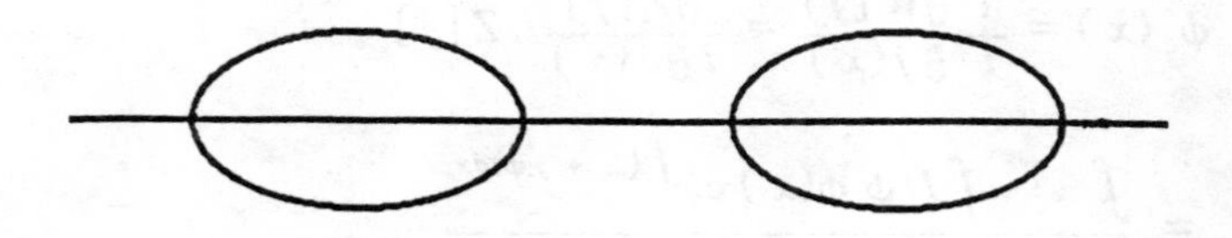

Fig.4-5a

Fig.4-5b 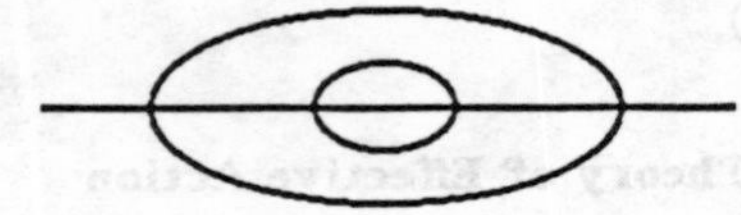

We define the proper function corresponding to a proper graph according to the Feynman rules without attaching the propagators to external lines. For example, the two point proper function corresponding to the graph (Fig. 4-5b) is given by

$$\frac{(-i\lambda)^2}{3!}(\Delta_F(x_1-x_2))^3 \tag{4.45}$$

We denote a general n-point proper function by $\kappa^{(n)}(x_1,x_2,\cdots,x_n)$ namely $\kappa^{(2)}(x_1,x_2)$ implies

$\kappa^{(2)}(x_1,x_2) =$ 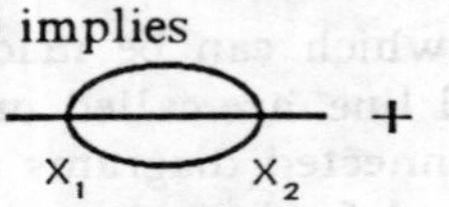

$+$ 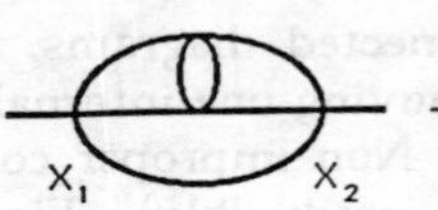

$+ \cdots\cdot$

$\equiv$ 

Next let us define $\phi_c(x)$ by

$$\phi_c(x) = \frac{1}{i}\frac{\partial W[j]}{\partial j(x)} = \frac{\partial Z[j]}{i\partial j(x)}/Z[j]$$

$$= \frac{\int\cdots\int D\phi\,\phi(x)\,e^{i\int(L+j\phi)dx}}{\int\cdots\int D\phi e^{i\int(L+j\phi)dx}}$$

$$\equiv \langle\phi(x)\rangle_j \tag{4.46}$$

which is an average of $\phi(x)$ in the presence of an external source function $j(x)$. Since $W[j]$ is the generating functional of connected Green's functions, i.e.

$$W[j] = \sum_n \frac{i^n}{n!}\int\cdots\int dx_1\cdots dx_n\, j(x_1)\ldots j(x_n)\, G_c^{(n)}(x_1,\cdots,x_n)$$

$$\equiv \sum_n \frac{i^n}{n!} \tag{4.47}$$ 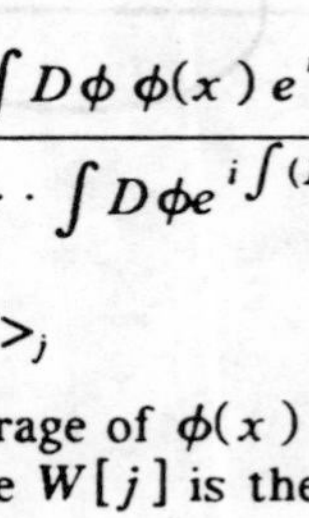

where $G_c^{(n)}(x_1,\ldots,x_n)$ is a n-point connected Green's function. $\phi_c(x)$ defined by (4.46) is a functional of $j(x)$:

$$\phi_c(x) = \frac{1}{i}\,\frac{\delta W[j]}{\delta j(x)} = \sum_n \frac{i^{n-1}}{(n-1)!}$$

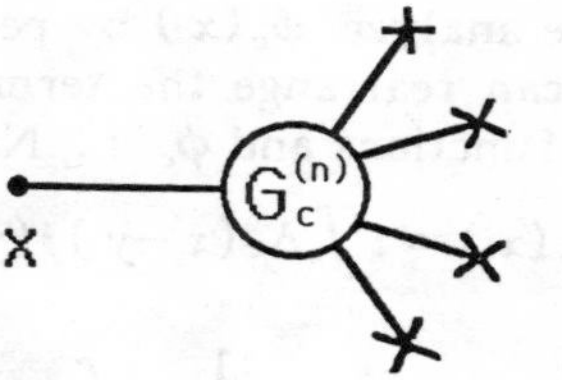

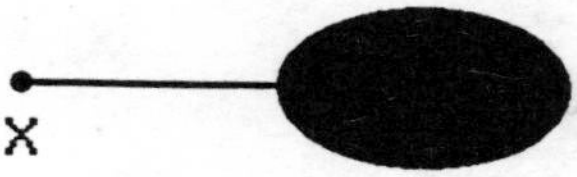

(4.48)

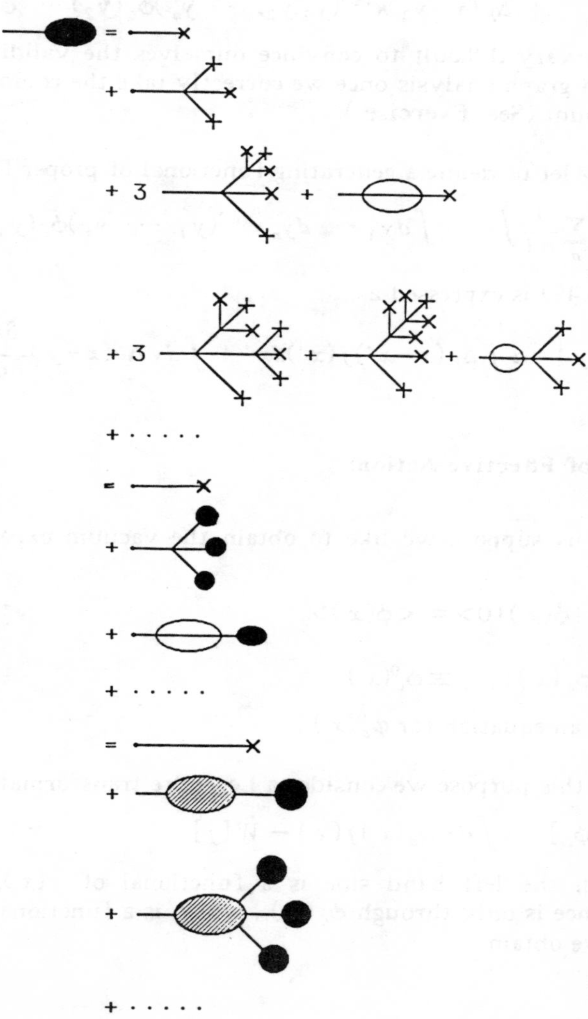

Fig. 4-6

We analyze $\phi_c(x)$ by perturbation expansion. As shown in Fig. 4-6, we can rearrange the terms so that $\phi_c(x)$ is expressed in terms of proper functions and ϕ_c's . Namely,

$$\phi_c(x) = i\int \Delta_F(x-y)j(y)dy +$$

$$+\sum_{n=2}\frac{1}{(n-1)!}\int\cdots\int dy_1\cdots dy_n$$

$$\Delta_F(x-y_1)\kappa^{(n)}(y_1,y_2,\cdots,y_n)\phi_c(y_2)\cdots\phi_c(y_n) \quad (4.49)$$

It is not very difficult to convince ourselves the validity of (4.49) by Feynman graph analysis once we correctly take the combinatorial factors into account (See **Exercise**).

Now let us define a generating functional of proper functions by

$$\kappa[\phi_c] = \sum_n \frac{1}{n!}\int\cdots\int dy_1\cdots dy_n\,\kappa^{(n)}(y_1,\cdots,y_n)\phi_c(y_1)\cdots\phi_c(y_n)$$

Then, (4.49) is expressed as

$$\phi_c(y) = i\int \Delta_F(x-x')j(x')dx' + \int dy\,\Delta_F(x-y)\frac{\delta\kappa[\phi_c]}{\delta\phi_c(y)} \quad (4.50)$$

Theory of Effective Action:

Let us suppose we like to obtain the vacuum expectation value of $\hat{\phi}(x)$:

$$<0|\hat{\phi}(x)|0> = <\phi(x)>_0$$

$$= \phi_c(x)|_{j=0} \equiv \phi_c^0(x) \quad (4.51)$$

We seek an equation for $\phi_c^0(x)$.

For this purpose we consider a Legendre transformation

$$i\Gamma[\phi_c] = i\int dx\,\phi_c(x)j(x) - W[j] \quad (4.52)$$

Although the left hand side is a functional of $j(x)$, the functional dependence is only through $\phi_c(x)$, which is a functional of $j(x)$. From (4.52) we obtain

$$\frac{\delta\Gamma[\phi_c]}{\delta\phi_c(x)} = j(x) \tag{4.53}$$

so that by setting $j(x) = 0$ we obtain

$$\frac{\delta\Gamma[\phi_c^0]}{\delta\phi_c^0(x)} = 0 \tag{4.54}$$

Namely we obtain an equation for ϕ_c^0 by making a variation of Γ with respect to ϕ_c^0. Thus, we call $\Gamma[\phi_c]$ an effective action.

By inserting (4.53) into (4.50) we obtain a functional differential equation for $\Gamma[\phi_c]$:

$$i\frac{\delta\Gamma[\phi_c]}{\delta\phi_c} = \int \Delta_F^{-1}(x-y)\phi_c(y)dy - \frac{\delta\kappa[\phi_c]}{\delta\phi_c(x)} \tag{4.55}$$

Thus,

$$i\,\Gamma[\phi_c] = \frac{1}{2}\int\int dxdy\,\phi_c(x)\Delta_F^{-1}(x-y)\phi_c(y) - \kappa[\phi_c] \tag{4.56}$$

Therefore, the effective action can be obtained from proper functions, which are calculable at least by perturbation.

V. Euclidean Field Theory and Statistical Mechanics

5.1 Statistical Mechanics, Euclidean Path Integral and Euclidean Field Theory

In quantum statistical mechanics one considers the partition function defined by

$$Z = tre^{-\beta\hat{H}} = \int dq <q \,|\, e^{-\beta\hat{H}} \,|\, q>, \quad \beta = (kT)^{-1} \tag{5.1}$$

where as usual k is the Boltzman constant and T is the temperature of the system. So, the relevant matrix element to consider is

$$<q \,|\, e^{-\beta\hat{H}} \,|\, q'> \tag{5.2}$$

which can be obtained from the Feynman kernel (2.2) by replacing $t-t'$ by $-i\beta$. This is accomplished by considering a complex t-plane and by rotating the positive real axis by 90 degree clockwise to the negative imaginary axis. This rotation is called the Wick rotation.

Since the Feynman kernel is expressed in terms of path integral (2.16), by a Wick rotation one obtains the path integral expression for (5.2) from (2.16). The action integral is then

$$i\int Ldt \rightarrow \int_0^\beta d\tau L(\,i\dot{q}(\tau), q(\tau)) \equiv -S \tag{5.3}$$

For example, for the following Minkowsky Lagrangian

$$L = \frac{1}{2}\dot{q}^2 - V(q) \tag{5.4}$$

one obtains

$$S = \int_0^\beta d\tau \left[\frac{1}{2}\dot{q}^2 + V(q)\right] \tag{5.5}$$

For a scalar field theory in d-dimensional space

$$L = \int d\vec{x} \left[\frac{1}{2}\dot{\phi}^2 - (\frac{1}{2}(\vec{\nabla}\phi)^2 + \frac{1}{2}m^2\phi^2 + ...)\right] \tag{5.6}$$

one obtains

$$S[\phi] \equiv S = \int d\vec{x} \int_0^\beta d\tau [\frac{1}{2}(\partial_\tau \phi)^2 + \frac{1}{2}(\vec{\partial}\phi)^2 + \frac{1}{2}m^2\phi^2 + ...] \quad (5.7)$$

which is the same as finite time Euclidean action.

The path integral expression for (5.1) is therefore given by

$$Z = \int_{q(0)=q(\beta)} \cdots \int Dq \; e^{-S} \quad (5.8)$$

Notice that the periodic boundary condition in (5.8) is its origin in the trace in the definition of partition function.

In statistical mechanics we often encounter a Hamiltonian of the system expressed in terms of creation and annihilation operators. Let us denote them by $\hat{a}^+$ and $\hat{a}$ both for Bose and Fermi. In this case we use holomorphic path integral of **III.** However, there should be one caution for the trace expression of Fermi operator. Namely, for a given operator $\hat{O}$ the trace is defined by

$$tr\hat{O} = <0|\hat{O}|0> + <1|\hat{O}|1>$$

$$= \int e^{-\bar{\psi}\psi} d\bar{\psi} d\psi <-\psi|\hat{O}|\psi> \quad (5.9)$$

which can be proven by using (3.39), (3.40) and the definition of Grassmann integration. Notice the sign difference between (5.1) and (5.9). Therefore, for given normal ordered Hamiltonian

$$\hat{H} = :H(\hat{a}^+, \hat{a}): \quad (5.10)$$

we obtain the following path integral expression of partition function by using (3.50) and (3.51) and the Wick rotation:

$$Z = \int_{\substack{\bar{\psi}(0)=\pm\bar{\psi}(\beta) \\ \psi(0)=\pm\psi(\beta)}} \cdots \int D\bar{\psi} D\psi e^{-S[\bar{\psi},\psi]} \quad (5.11)$$

where

$$S[\bar{\psi},\psi] = \int_0^\beta d\tau [\bar{\psi}(\tau)\dot{\psi}(\tau) + H(\bar{\psi},\psi)] \quad (5.12)$$

In (5.11) + sign is for Bose and - sign for Fermi. It is remarkable that the appearance of path integral for Bose and Fermi system is almost identical except the sign difference in the boundary condition (periodic for Bose and anti-periodic for Fermi), although the nature of the integrations is so different, i.e. ordinary and Grassmannian.

5.2 Perturbation Expansion

Let $\hat{H}$ be given by

$$\hat{H} = \hat{H}_0 + H_1(\hat{a}^+, \hat{a}): \tag{5.13}$$

$$\hat{H}_0 = \omega \hat{a}^+ \hat{a} \tag{5.14}$$

Since the path integral for Z is similar, we can apply the same perturbation expansion technique as in **IV**. However, there are some tricky sign problems in the process, so we repeat some of the familiar procedures.

Let $Z_0[\bar{\eta}, \eta]$ be a free partition function with external sources:

$$Z_0[\bar{\eta}, \eta] = \int_{\substack{\psi(0)=\pm\psi(\beta) \\ \bar{\psi}(0)=\pm\bar{\psi}(\beta)}} \cdots \int D\bar{\psi} D\psi e^{-\int_0^\beta (\bar{\psi}\dot{\psi} + \omega\bar{\psi}\psi - \bar{\eta}\psi - \bar{\psi}\eta) d\tau} \tag{5.15}$$

The upper sign is for Bose and lower for Fermi. We also use Grassmann variables for η and $\bar{\eta}$ in the case of Fermi. Then,

$$Z = e^{-\int_0^\beta d\tau H_1 \left[\pm \frac{\delta}{\delta \eta}, \frac{\delta}{\delta \bar{\eta}}\right]} Z_0[\bar{\eta}, \eta]\Big|_{\eta=\bar{\eta}=0} \tag{5.16}$$

We calculate Z_0 by shifting ψ and $\bar{\psi}$ as before:

$$\psi(\tau) \to \psi(\tau) + \psi_0(\tau), \qquad \bar{\psi}(\tau) \to \bar{\psi}(\tau) + \bar{\psi}_0(\tau) \tag{5.17}$$

The exponent (5.15) is then given by

$$-\int_0^\beta (\bar{\psi}\dot{\psi} + \omega\bar{\psi}\psi) d\tau - \int_0^\beta (\bar{\psi}_0\dot{\psi}_0 + \omega\bar{\psi}_0\psi_0 - \bar{\eta}\psi_0 - \bar{\psi}_0\eta) d\tau -$$

$$- \int_0^\beta (\bar{\psi}\dot{\psi}_0 + \omega\bar{\psi}\psi_0 - \bar{\psi}\eta) d\tau - \int_0^\beta (\bar{\psi}_0\dot{\psi} + \omega\bar{\psi}_0\psi - \bar{\eta}\psi) d\tau$$

Therefore, to cancel the second line we require

$$(\partial_\tau + \omega)\psi_0 = \eta, \qquad (-\partial_\tau + \omega)\bar{\psi}_0 = \bar{\eta} \tag{5.18}$$

with boundary conditions

$$\psi_0(0) = \pm\psi_0(\beta), \qquad \bar{\psi}_0(0) = \pm\bar{\psi}_0(\beta) \tag{5.19}$$

Let us define Matsubara Green's function by

$$(\frac{\partial}{\partial \tau} + \omega)\Delta(\tau,\tau') = \delta(\tau-\tau'), \qquad \beta \geqslant \tau,\tau' \geqslant 0 \tag{5.20}$$

$$\Delta(0,\tau') = \pm\,\Delta(\beta,\tau') \tag{5.21}$$

Use

$$\sum_{m=-\infty}^{\infty} \delta(\tau-\tau' + m\beta) = \frac{1}{\beta}\sum_{n=-\infty}^{\infty} e^{-\frac{2\pi i n}{\beta}(\tau-\tau')} \tag{5.22}$$

to obtain (see **Exercise** .)

$$\Delta(\tau,\tau') = \frac{1}{\beta}\sum_{n=-\infty}^{\infty}\frac{1}{\omega-i\xi_n}\,e^{-i\xi_n(\tau-\tau')} = \frac{1}{\beta}\sum_{n}\Delta_n e^{-i\xi_n(\tau-\tau')} \tag{5.23}$$

where

$$\xi_n = \frac{\pi}{\beta}2n \qquad \text{for} \quad \textit{Bose}$$

$$\xi_n = \frac{\pi}{\beta}(2n+1) \qquad \text{for} \quad \textit{Fermi} \tag{5.24}$$

We note

$$\Delta^*(\tau,\tau') = \Delta(\tau',\tau)$$

$$(-\frac{\partial}{\partial \tau'} + \omega)\Delta(\tau,\tau') = \delta(\tau-\tau') \tag{5.25}$$

Thus, the solution of $\psi_0(\tau)$ and $\bar{\psi}_0(\tau)$ is given by

$$\psi_0(\tau) = \int_0^\beta d\tau'\Delta(\tau,\tau')\eta(\tau')$$

$$\bar{\psi}_0(\tau) = \int_0^\beta d\tau'\bar{\eta}(\tau')\Delta(\tau',\tau) \tag{5.26}$$

It is simple now to obtain

$$Z_0[\bar{\eta},\eta] = e^{\int_0^\beta d\tau\int_0^\beta d\tau'\bar{\eta}(\tau)\Delta(\tau,\tau')\eta(\tau')}\, Z[0,0] \tag{5.27}$$

Inserting (5.27) into (5.16) and using (4.33) we obtain

$$Z[\bar{\eta},\eta] = e^{\pm\int_0^\beta d\tau \int_0^\beta d\tau' \Delta(\tau,\tau') \frac{\partial}{\partial\psi(\tau)} \frac{\partial}{\partial\bar{\psi}(\tau')}}$$

$$e^{-\int_0^\beta d\tau [H_1(\bar{\psi},\psi) - \bar{\psi}\eta - \bar{\eta}\psi]} \Big|_{\psi=\bar{\psi}=0} \qquad (5.28)$$

Now the perturbation expansion is done straightforwardly. We show it in the next section for the BCS Hamiltonian.

Exercise: Obtain (5.23).

5.3 Application to BCS Theory of Superconductivity

Feynman Rules:

The Hamiltonian for the superconducting system (BCS Hamiltonian) is given by

$$\hat{H} = \hat{H}_0 + \hat{H}_1$$

$$\hat{H}_0 = \sum_{s=\uparrow\downarrow} \int d\vec{x}\, \hat{\psi}_s^+(\vec{x})(- \vec{\nabla}\frac{2}{2m} - \mu)\, \hat{\psi}_s(\vec{x})$$

$$\hat{H}_1 = -g \int d\vec{x}\, \hat{\psi}_\uparrow^+(\vec{x}) \hat{\psi}_\downarrow^+(\vec{x}) \hat{\psi}_\downarrow(\vec{x}) \hat{\psi}_\uparrow(\vec{x}) \qquad (5.29)$$

where g is a positive coupling constant (attractive) between spin up ($\uparrow$) and spin down ($\downarrow$) electron and μ is the chemical potential. We can introduce electro-magnetic interaction in a gauge-invariant way by the minimal substitution

$$\vec{\nabla} \to \vec{\nabla} - ie\vec{A}$$

and treat the vector potential $\vec{A}(\vec{x})$ as an external source. Then the Hamiltonian

$$\hat{H} = \sum_{s=\uparrow,\downarrow} \int d\vec{x}\, \hat{\psi}_s^+(\vec{x})[- \frac{1}{2m}(\vec{\nabla} - ie\vec{A})^2 - \mu]\hat{\psi}_s(\vec{x})$$

$$- g \int d\vec{x}\, \hat{\psi}_\uparrow^+(\vec{x}) \hat{\psi}_\downarrow^+(\vec{x}) \hat{\psi}_\downarrow(\vec{x}) \hat{\psi}_\uparrow(\vec{x}) \qquad (5.30)$$

is invariant under gauge transformation

$$\psi_s(\vec{x}) \to e^{ie\Lambda(\vec{x})}\psi_s(\vec{x}), \qquad \vec{A}(\vec{x}) \to \vec{A}(\vec{x}) + \vec{\nabla}\Lambda(\vec{x}) \tag{5.31}$$

The partition function is given by

$$Z = \int D\psi D\bar{\psi} \exp\left[-\int_0^\beta d\tau \int_V d\vec{x}\,\bar{\psi}_s\left[\frac{\partial}{\partial\tau} - \frac{1}{2m}(\vec{\nabla} - ie\vec{A})^2 - \mu\right]\psi_s\right]$$

$$\exp\left[g\int_0^\beta d\tau \int_V d\vec{x}\,\bar{\psi}_\uparrow(\vec{x})\bar{\psi}_\downarrow(\vec{x})\psi_\downarrow(\vec{x})\psi_\uparrow(\vec{x})\right] \tag{5.32}$$

The 4-fermion interaction can be expressed in terms of a complex auxiliary scalar field as

$$\exp[-\int_0^\beta d\tau\, H_1(\bar{\psi},\psi)] = \frac{1}{C}\int D\phi D\phi^*$$

$$\exp[-\kappa^2\int d^4x\,\phi^*\phi + g^{1/2}\kappa\int d^4x(\bar{\psi}_\uparrow\bar{\psi}_\downarrow\phi + \psi_\downarrow\psi_\uparrow\phi^*)] \tag{5.33}$$

where we have used the notation

$$\int d^4x \equiv \int_0^\beta d\tau \int_V d^3\vec{x} \tag{5.34}$$

The constant C is given by

$$C = \int D\phi D\phi^* \exp[-\kappa^2\int d^4x\,\phi^*\phi] \tag{5.35}$$

where κ is a constant with dimension of mass, which we put to have the correct dimension for ϕ. Of course the physical result should be independent of κ. We shall introduce a source for the ϕ field as

$$Z[j,j^*] = \frac{1}{C}\int D\psi D\bar{\psi} D\phi D\phi^* \exp\left[-\int d^4x\,\bar{\psi}_s\left[\frac{\partial}{\partial\tau} - \frac{(\vec{\nabla} - ie\vec{A})^2}{2m} - \mu\right]\psi_s\right]$$

$$\exp[-\kappa^2\int\phi^*\phi d^4x + g^{1/2}\kappa\int d^4x(\bar{\psi}_\uparrow\bar{\psi}_\downarrow\phi + \psi_\downarrow\psi_\uparrow\phi^* + j^*\phi + \phi^* j)] \tag{5.36}$$

The partition function is

$$Z = Z[j,j^*]|_{j=j^*=0} \tag{5.37}$$

The Lagrangian appeared in (5.36) is

$$L = L_0 + L_1 \tag{5.38}$$

$$L_0 = \int d\vec{x} \left[\bar{\psi}_s \left(\frac{\partial}{\partial \tau} - \frac{\vec{\nabla}^2}{2m} - \mu \right) \psi_s + \kappa^2 \phi^* \phi \right] \tag{5.39}$$

$$L_1 = \int d\vec{x} \left[-\frac{ie}{2m} (\bar{\psi}_s \vec{\nabla} \psi_s - (\vec{\nabla} \bar{\psi}_s) \psi_s) \cdot \vec{A} + \frac{e^2}{2m} \bar{\psi}_s \psi_s \vec{A}^2 + \right.$$
$$\left. + g^{1/2} \kappa (\bar{\psi}_\uparrow \bar{\psi}_\downarrow \phi + \psi_\downarrow \psi_\uparrow \phi^*) \right] \tag{5.40}$$

Using the Fourier decomposition

$$\psi_s(\vec{x},\tau) = (\beta V)^{-1/2} \sum_{k} \tilde{\psi}_s(\vec{k},n)\, e^{i(\vec{k}\cdot\vec{x} - \xi_n \tau)} \tag{5.41}$$

we obtain the free action

$$\int_0^\beta d\tau L_0 = \sum_{\vec{k},n} \left[\tilde{\bar{\psi}}_s(\vec{k},n)(-i\,\xi_n + \omega_k)\tilde{\psi}_s(\vec{k},n) + \kappa^2 \tilde{\phi}^*(\vec{k},n)\tilde{\phi}(\vec{k},n) \right] \tag{5.42}$$

Therefore, the propagator of electron with four momentum $(\vec{k},n)$ is given by

$$\frac{1}{\omega_k - i\,\xi_n}, \quad \omega_k = \frac{\vec{k}^2}{2m} - \mu, \quad \xi_n = \frac{\pi}{\beta}(2n+1) \tag{5.43}$$

and the propagator of ϕ by

$$\frac{1}{\kappa^2} \tag{5.44}$$

Notice that the propagator of ϕ does not depend on four momentum, because Lagrangian (5.39) does not contain the kinetic energy term.

By substituting Fourier decomposition of the fields into the interaction Lagrangian (5.40), we can easily read off the Feynman rules:

propagator $= 1/\kappa^2$ ϕ

propagator $= (\omega_p - i\xi)^{-1}$ $\Psi_\uparrow$ p=($\vec{p}$,n)

propagator $= (\omega_p - i\xi)^{-1}$ $\Psi_\downarrow$ p=($\vec{p}$,n)

vertex $= (g/\beta V)^{1/2}\kappa$

vertex $= (g/\beta V)^{1/2}\kappa$

vertex $= \dfrac{e}{2m}(\vec{p}\,' + \vec{p})$

vertex $= e^2/2m$

Derivation of Landau-Ginzburg Equation:

The generating functional for the connected Green's functions is defined as

$$W[j, j^*] = \ln Z[j, j^*] \tag{5.45}$$

The Legendre transformation of $W[j, j^*]$ is

$$\Gamma[\phi_c, \phi^*_c; \vec{A}] = \int d^4x\,(j\phi^*_c + \phi_c j^*) - W[j, j^*] \tag{5.46}$$

The partition function is

$$Z = Z[j, j^*]_{j=j^*=0} = \exp\left(-\Gamma[\phi_c, \phi^*_c; \vec{A}]|_{j=j^*=0}\right) \tag{5.47}$$

which implies that Γ has to satisfy

$$\left.\frac{\delta\Gamma}{\delta\phi_c(x)}\right|_{j=j^*=0} = 0, \qquad \left.\frac{\delta\Gamma}{\delta\phi^*_c(x)}\right|_{j=j^*=0} = 0 \tag{5.48}$$

The generating functional for the one-particle-irreducible diagrams is given by

$$\kappa[\phi_c, \phi^*_c; \vec{A}] = \int d^4x\, \mu^2 \phi^*_c(x) \phi_c(x) - \Gamma[\phi, \phi^*; \vec{A}] \qquad (5.49)$$

Calculation of 1-loop diagrams:

We shall first compute all 1-loop diagrams with 2 and 4 external ϕ - lines. The term proportional to $|\phi_c|^2$ in $\Gamma[\phi_c, \phi^*_c]$ is

p

$q_o = 0$

q q

q−p=p'

$$= \frac{g\kappa^2}{\beta V} \sum_{n,\vec{p}} \frac{1}{(i\xi_n - \omega_p)(-i\xi_n - \omega_{p'})}$$

$$= \frac{g\kappa^2}{\beta V} \sum_{n,\vec{p}} \frac{1}{(i\xi_n - \omega_p)} e^{-\vec{q}\cdot\frac{\partial}{\partial\vec{p}}} \frac{1}{(-i\xi_n - \omega_p)}$$

$$= A + B\vec{q}^2 + \ldots\ldots \qquad (5.50)$$

where

$$A = \frac{g\kappa^2}{\beta V} \sum_{n,\vec{p}} \frac{1}{\xi_n^2 + \omega_p^2}$$

$$= \sum_n \frac{g\kappa^2}{\beta} \frac{1}{(2\pi)^3} \int d^3\vec{p}\, \frac{1}{\xi_n^2 + \omega_p^2} \qquad (5.51)$$

Introducing density of states $\rho(\omega)$ as

$$\frac{1}{(2\pi)^3} \int d^3\vec{p} \equiv \int d\omega_p\, \rho(\omega_p) \qquad (5.52)$$

we get

$$A = \sum_n \frac{g\kappa^2}{\beta} \int_{-\infty}^{\infty} \frac{d\omega\, \rho(\omega)}{\xi_n^2 + \omega^2}$$

$$\approx \sum_n \frac{g\kappa^2}{\beta} \rho(0) \int_{-\infty}^{\infty} \frac{d\omega}{\xi_n^2 + \omega^2}$$

$$= 2g\kappa^2\rho(0)\sum_{n=0}^{n_{\max}} \frac{1}{2n+1} \tag{5.53}$$

where $\omega_p = \dfrac{p^2}{2m} - \mu = 0$ is the Fermi surface and we have approximated by using density of states at the Fermi surface and also put in a cut-off $n_{\max}$ which is equivalent to a cut-off in energy (Debye energy) equal to

$$\omega_D = kT\pi(2n_{\max} + 1) \tag{5.54}$$

$$A = g\kappa^2\rho(0)\ln\frac{4\gamma\omega_D}{\pi kT}, \qquad \gamma = \textit{Euler constant} \tag{5.55}$$

The constant B is given by

$$B = \frac{1}{q^2}\frac{g\kappa^2}{\beta V}\sum_{n\vec{p}} \frac{1}{i\xi_n - \omega_p}\,\frac{1}{2}\left(\vec{q}\cdot\frac{\partial}{\partial\vec{p}}\right)^2\frac{1}{-i\xi_n - \omega_p}$$

$$= \frac{1}{6}\frac{g\kappa^2}{\beta V}\sum_{n\vec{p}} \frac{1}{i\xi_n - \omega_p}\,\frac{\partial^2}{\partial\vec{p}^2}\,\frac{1}{-i\xi_n - \omega_p}$$

$$= \frac{g\kappa^2}{6\beta m V}\sum_{n\vec{p}} - \frac{2(\mu + \omega_p)}{(\xi_n^2 + \omega_p^2)^2}$$

$$= -\frac{g\kappa^2\rho(0)}{6\beta m}\sum_n \int_{-\infty}^{\infty} \frac{2\mu\, d\omega}{(\xi_n^2 + \omega^2)^2}$$

$$= -\frac{g\kappa^2\mu\rho(0)}{3\beta m}\sum_n \frac{\pi}{2}\,\frac{1}{|\xi_n|^3} \tag{5.56}$$

Using

$$\sum_{n=-\infty}^{\infty} \frac{1}{|2n+1|^3} = 2\sum_{n=0}^{\infty} \frac{1}{|2n+1|^3} = \frac{1}{4}\sum_{n=0}^{\infty} \frac{1}{|n+½|^3}$$

$$= \frac{1}{4}\zeta(3, 1/2) \tag{5.57}$$

Note

$$\sum_0^{\infty} \frac{1}{(n+b)^a} = \zeta(a, b)$$

$$B = -\frac{g\kappa^2\mu\rho(0)}{24m}\zeta(3,1/2)\frac{1}{(\pi kT)^2} \tag{5.58}$$

The term proportional to $(\phi_c{}^*\phi_c)^2$ in $\Gamma[\phi_c, \phi_c{}^*]$ is

$$= -\frac{g^2\kappa^4}{(\beta V)^2}\sum_{n,\vec{p}}\frac{1}{(\xi_n^2+\omega_p^2)^2}$$

$$= -\frac{g^2\kappa^4\rho(0)}{\beta(\beta V)^2}\frac{\beta^3}{8\pi^2}\zeta(3,1/2) \tag{5.59}$$

The coefficient of $|\phi_c|^2$ term in $\Gamma[\phi_c, \phi_c{}^*]$ is

$$\kappa^2\left(1 - g\rho(0)\ln\frac{4\gamma\omega_D}{\pi kT}\right)\phi_c{}^*\phi_c \tag{5.60}$$

for which the critical temperature is

$$T_c = \frac{4\gamma\omega_D}{\pi k}\exp\left(-\frac{1}{g\rho(0)}\right) \tag{5.61}$$

which is independent of κ. Thus, the $\phi_c{}^*\phi_c$ term is

$$\kappa^2\left(1 - g\rho(0)\ln\left[\frac{T_c}{T}\exp\frac{1}{g\rho(0)}\right]\right)\phi_c{}^*\phi_c =$$

$$= \left(\kappa^2 g\rho(0)\ln\frac{T}{T_c}\right)\phi_c{}^*\phi_c \tag{5.62}$$

To write down the complete gauge invariant effective action to 1-loop, we also have to compute the $\phi_c{}^*\phi_c\vec{A}$ and $\phi_c{}^*\phi_c\vec{A}^2$ graphs which are as follows. (Although the results are obtained from (5.50) by a minimum substitution, we compute these graphs in order to make sure for the gauge invariance of Γ.):

i) Coefficient of $\phi^*\phi A_i$ is given by

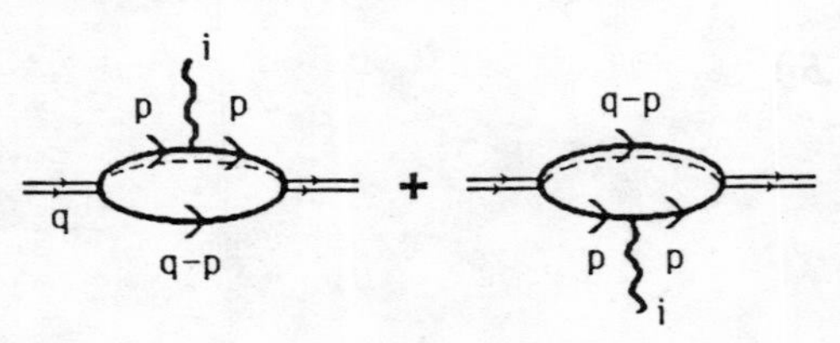

We are only interested in terms linear in q. We chose $q_0 = 0$ and the above graphs are

$$2\frac{g\kappa^2}{\beta V}\left(\frac{-ie}{2m}\right)\sum_{n,\vec{p}}\frac{2ip_i}{(i\xi_n-\omega_p)^2}\frac{1}{(-i\xi_n-\omega_{\vec{p}-\vec{q}})}$$

$$= 2\frac{g\kappa^2}{\beta V}\left(\frac{-ie}{2m}\right)\sum_{n,\vec{p}}\frac{2ip_i}{(i\xi_n-\omega_p)^2}e^{-\vec{q}\cdot\frac{\partial}{\partial\vec{p}}}\frac{1}{(-i\xi_n-\omega_p)} \qquad (5.63)$$

Using

$$\int d\,\Omega_{\vec{p}}\, p_i p_j = \frac{1}{3}\,\delta_{ij}\,\vec{p}^{\,2} \qquad (5.64)$$

we get

$$\frac{2g\kappa^2\mu e\rho(0)}{3\beta m}q_i\frac{\beta^3}{\pi^2}\frac{1}{4}\zeta(3,1/2)$$

$$=\frac{g\kappa^2\mu e\rho(0)}{6m}\frac{q_i}{(\pi kT)^2}\zeta(3,1/2) \qquad (5.65)$$

ii) Coefficient of $\phi_c^*\phi_c A_i A_j$ is given by

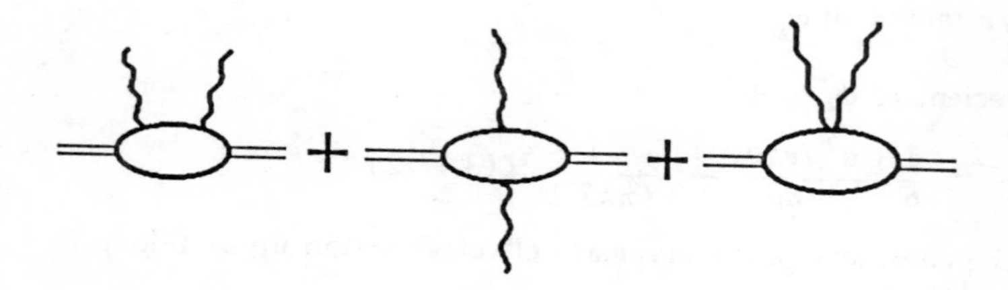

We are only interested in graphs having all external momenta equal to zero.

$$First\ graph = 2\frac{g\kappa^2}{\beta V}\left(\frac{-ie}{2m}\right)^2\sum_{n,\vec{p}}\frac{(2ip_i)(2ip_j)}{(i\xi_n-\omega_p)^3(-i\xi_n-\omega_p)}$$

$$= \frac{2g\kappa^2 e^2}{4\beta m^2} \; 8m\mu\rho(0)(-\frac{x}{2})\frac{1}{3}\delta_{ij} \tag{5.66}$$

where

$$x = \sum_n \frac{\pi}{2} \frac{1}{|\xi_n|^3} \tag{5.67}$$

$$\textit{Second graph} = \frac{\kappa^2 g}{\beta V}(\frac{-ie}{2m})^2 \sum_{n\,\vec{p}} \frac{(2ip_i)(-2ip_j)}{(i\xi_n - \omega_p)^2(-i\xi_n - \omega_p)^2}$$

$$= -\frac{2e^2\kappa^2 g\rho(0)}{\beta m}(x)\frac{1}{3}\delta_{ij} \tag{5.68}$$

$$\textit{Third graph} = 2\frac{e^2}{2m}\frac{g\kappa^2}{\beta V}\sum_{n\,\vec{p}} \frac{\delta_{ij}}{(i\xi_n - \omega_{\vec{p}})^2(-i\xi_n - \omega_p)}$$

$$= \frac{e^2 g\kappa^2}{\beta m V}\sum_{n\,\vec{p}} \frac{-i\xi_n - \omega_p}{(\omega_p^2 + \xi_n^2)^2}\delta_{ij}$$

$$= 0 \tag{5.69}$$

since first term in numerator is anti-symmetric in n and 2nd term is anti-symmetric in ω_p .

So coffecient of $\phi_c^*\phi_c\vec{A}^2$

$$= -\frac{1}{6}\frac{g\kappa^2\mu e^2\rho(0)}{m}\frac{1}{(\pi kT)^2}\zeta(3,1/2) \tag{5.70}$$

Then the complete gauge-invariant effective action up to 1-loop is

$$\Gamma[\phi_c, \phi_c^*; \vec{A}] = a\phi_c^*\phi_c + b(\phi_c^*\phi_c)^2$$

$$+ c\phi_c^*[-\frac{1}{4m}(\vec{\nabla} - 2ie\vec{A})^2]\phi_c + \frac{1}{2}(\vec{\nabla}\times\vec{A})^2 \tag{5.71}$$

where

$$a = \kappa^2 g\rho(0)\ln\frac{T}{T_c}$$

$$b = \frac{1}{8}g^2\kappa^4\rho(0)\zeta(3,1/2)\frac{1}{(\pi kT)^2}$$

$$c = \frac{1}{6} g\kappa^2 \mu\rho(0) \frac{1}{(\pi kT)^2} \zeta(3,1/2) \tag{5.72}$$

It is worthwhile to note that ϕ acts like a composite particle of mass 2m and change 2e.

The condition of minimum of effective action gives the Landau-Ginzburg equation as

$$\frac{\delta\Gamma}{\delta\phi^*} = 0 = a\phi + 2b\,|\phi|^2\phi + c[-\frac{1}{4m}(\vec{\nabla} - 2ie\vec{A})^2]\phi \tag{5.73}$$

Assuming a space-independent ϕ, we get for the minimum

$$|\phi_{min}|^2 = -a/2b \tag{5.74}$$

Remarks:

The derivation of Landau-Ginzburg equation from BCS theory was first done by Gorkov by using the Green function method. The calculation presented here is essentially the same as Gorkov's in its content, but the use of effective action illuminates more the justification of Landau-Ginzburg phenomenological theory.

Higgs Mechanism:

The Higgs model is defined by the following Lagrangian

$$L = -\frac{1}{4}F_{\mu\nu}F^{\mu\nu} + |(\partial_\mu - ieA_\mu)\phi|^2 - V(\phi) \tag{5.75}$$

$$V(\phi) = a\,|\phi|^2 + b(|\phi|^2)^2$$

$$F_{\mu\nu} = \partial_\mu A_\nu - \partial_\nu A_\mu \tag{5.76}$$

If we ignore the time dependence, this model is identical to the Landau-Ginzburg model,(5.71). Namely the Higgs model is a relativistic version of Landau-Ginzburg model, so here we study the Higgs model instead.

If a is negative $V(\phi)$ has a minimum at

$$|\phi|^2 = \lambda^2, \qquad \lambda = (|a|/2b)^{1/2} \tag{5.77}$$

Let us parameterize ϕ field by

$$\phi(x) = e^{i\chi(x)/2^{1/2}\lambda}\left[\lambda + 2^{-1/2}\eta(x)\right] \tag{5.78}$$

then we obtain

$$|\partial_\mu\phi|^2 - V(\phi) = (\partial_\mu\chi)^2 + \frac{1}{2}(\partial_\mu\eta)^2 + \frac{m_S^2}{2}\eta^2 + \cdots \tag{5.79}$$

where

$$m_S = (2|a|)^{1/2} \tag{5.80}$$

$\chi(x)$ is massless field, but $\eta(x)$ represents a massive field with mass given by (5.80). Since $\chi(x)$ is the phase of $\phi(x)$, $V(\phi)$ is invariant by the change of $\chi(x)$, accordingly $\chi(x)$ is massless field (Nambu-Goldstone boson).

In the presence of electromagnetic field, however, the Lagrangian is expressed in terms of $\chi(x)$ and $\eta(x)$ as

$$L = -\frac{1}{4}F_{\mu\nu}F^{\mu\nu} - \frac{1}{2}(\partial_\mu\chi)^2 - \frac{1}{2}(\partial_\mu\eta)^2 - \frac{m_S^2}{2}\eta^2 - e^2\lambda^2A_\mu^2 + 2e\lambda A_\mu\partial^\mu\chi + \cdots \tag{5.81}$$

so that if we define

$$A_\mu' = A_\mu - (2^{1/2}e\lambda)^{-1/2}\partial_\mu\chi \tag{5.82}$$

the Lagrangian is expressed entirely in terms of $\eta(x)$ and $A_\mu'(x)$ and $\chi(x)$ field disappears. We use $A_\mu(x)$ for $A_\mu'(x)$:

$$L = -F_{\mu\nu}F^{\mu\nu} - \frac{1}{2}(\partial_\mu\eta)^2 - \frac{m_S^2}{2}\eta^2 - \frac{m_V^2}{2}A_\mu^2 + \cdots \tag{5.83}$$

where

$$m_V = 2^{1/2}e\lambda \tag{5.84}$$

The gauge field A_μ acquires the mass m_V. This mechanism, that the massless vector field becomes massive in terms of the spontaneous symmetry breaking, is called the Higgs mechanism. Originally, the photon field has two independent transverse components and there are two scalar bosons. After the Higgs mechanism, the photon acquires the mass so that it has a longitudinal component and only one scalar meson left.

The Maxwell equation is modified accordingly:

$$\partial^\mu F_{\mu\nu} + m_V^2 A_\nu = 0, \tag{5.85}$$

which can be written in the static case (i. e $\dot{A} = 0$)

$$\nabla^2 \vec{A} - \vec{\nabla}(\vec{\nabla}\cdot\vec{A}) + m_V^2 \vec{A} = 0 \tag{5.86}$$

By taking curl of (5.86) we obtain the equation for magnetic field $\vec{H} = \vec{\nabla}\times\vec{A}$.

$$(\nabla^2 + m_V^2)\vec{H} = 0 \tag{5.87}$$

$\vec{H} = 0$ is the only constant solution (uniform i.e. independent of x). Therefore, in a uniform superconductor there is no magnetic field (Meisner effect).

The Meisner effect can be derived also by the following simple consideration. In the superconductor state ($a < 0$) the vacuum expectation value of $\phi(x)$ is a non-zero canstant. Thus,

$$\vec{\nabla} < \phi(x) >_0 = 0 \tag{5.88}$$

Since $\phi(x)$ is a charged scalar field, (note in BCS theory $\phi(x)$ is the field of Cooper pair (see (5.33)) and an order parameter of superconductor), due to the gauge invariance the equation (5.88) should be modified as

$$\left(\vec{\nabla} - ie\vec{A}(x)\right) < \phi(x) >_0 \equiv \vec{D} < \phi(x) >_0 = 0 \tag{5.89}$$

Using

$$[D_i , D_j] = -ieF_{ij} = -ie\,\epsilon_{ijk} H_k \tag{5.90}$$

and (5.89) one obtains

$$\vec{H}(x) < \phi(x) >_0 = 0 \tag{5.91}$$

In the superconductor state $< \phi >_0 \neq 0$ so one obtains the Meisner effect.

Next let us consider the vicinity of flat surface of superconductor. Let $\hat{x}$ be the normal direction to the surface and $x = 0$ be the position of the surface. Let

$$\phi(\vec{x},t) = \lambda f(x) \tag{5.92}$$

namely a function of x only. Then the Landau-Ginzburg equation becomes

$$-\xi^2 \frac{d^2}{dx^2} f - f^3 = 0 \tag{5.93}$$

where ξ is a constant, called the coherent length.

$$\xi^2 = \frac{1}{m_S^2} = \frac{1}{2\,|\,a\,|} \tag{5.94}$$

Then we should impose the following boundary conditions:

$$f\,(\infty) = 1 \qquad (\textit{inside of superconductor})$$

$$f\,(x) = 0 \quad \text{for } x < 0 \qquad (\textit{outside of superconductor}) \tag{5.95}$$

Multiply x to (5.89) and integrate and then use boundary condition (5.91) to fix the integration constant. We obtain

$$-\,\xi^2 \left(\frac{df}{dx} \right)^2 + \frac{1}{2}(1 - f^{\,2})^2 = 0 \tag{5.96}$$

$$\frac{df}{dx} = \frac{1}{2^{1/2}\xi}(1 - f^{\,2}), \quad \text{and} \quad f\,(x) = \tanh(x\,/\,2^{1/2}\xi) \tag{5.97}$$

When a magnetic field is applied we get a solution as shown in Fig. 5-1.

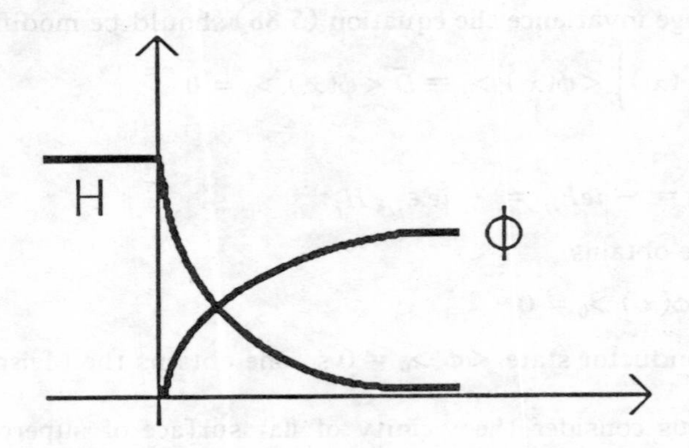

Fig. 5-1

Near the surface $x = 0$, there is a skin effect. The penetration length is $\xi = 1/\,m_S$ for f and $1/\,m_V$ for H. The superconductors which satisfy

$$m_S < m_V \quad (type I)$$

is called "type I" , while

$$m_S > m_V \quad (type II)$$

In type II superconductors the magnetic field can penetrate up to the length $\sim 1/m_V$ without destroying the superconductor state, i.e. $f \sim 1$.

Abrikosov-Nielsen-Olesen Vortex Solution:

Let us start with the Abelian Higgs model Lagrangian (5.75) and (5.76) with *negative* a. The equations of motion derived from (5.75) are

$$(\partial_\mu - ieA_\mu)^2 \phi + |a| \phi - 2b |\phi|^2 \phi = 0 \tag{5.98}$$

$$\partial^\nu F_{\mu\nu} = j_\mu = 2e^2 |\phi|^2 A_\mu + ie(\phi^* \partial_\mu \phi - \phi \partial_\mu \phi^*) \tag{5.99}$$

With the parametrization of complex field

$$\phi(x) = |\phi| e^{i\chi}$$

the vector potential is expressed as

$$eA_\mu = \frac{j_\mu}{2} e |\phi|^2 + \partial_\mu \chi \tag{5.100}$$

Let us consider a region of space where $j_\mu = 0$ and a contour C in this region.

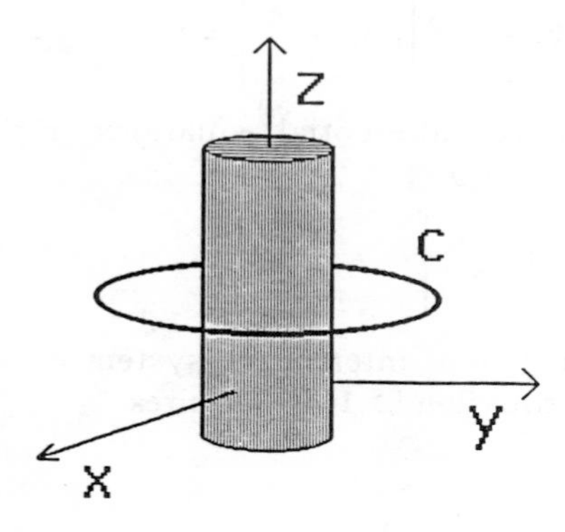

Fig. 5-2

Using (5.100) and performing a line integral along the contour, we get

$$\int_C eA_\mu dx^\mu = \int_C (\partial_\mu \chi) dx^\mu \tag{5.101}$$

The line integral over the gradient of the phase of ϕ does not necessarily vanish. The only general requirement on the phase is that ϕ is single valued, i.e., χ varies by $2\pi n$ (n = integer) when we make a complete turn around a closed loop. So, we get from (5.101)

$$e\int_C A_\mu dx^\mu = e\int_S F_{\mu\nu} d\,\sigma^{\mu\nu} = 2\pi n \tag{5.103}$$

where S is an area of which C is the boundary. Since the left hand side of (5.103) is the magnetic flux, this solution brings "quantized" magnetic flux.

Let us take a cylindrical coordinates, which is natural for the configuration of Fig. 5-2, and seek a solution with the following ansatz:

$$A_0 = A_z = A_r = 0, \qquad A_\theta = A(r), \qquad |\phi| = g(r), \qquad \chi = 2\pi n\,\theta \tag{5.104}$$

The magnetic field is then in z-direction and its strength is given by

$$H = -\frac{1}{r}\frac{d}{dr}(rA) \tag{5.105}$$

The equations of motion,(5.98) and (5.99), are

$$\frac{1}{r}\frac{d}{dr}\left(r\frac{d}{dr}g\right) + \left[\left(\frac{n}{r} - eA\right)^2 + |a|^2 - 2bg^2\right] g = 0 \tag{5.106}$$

$$-\frac{d}{dr}\left(\frac{1}{r}\frac{d}{dr}rA\right) + eg^2\left(eA - \frac{n}{r}\right) = 0 \tag{5.107}$$

We solve these coupled differential equations imposing the following asymptotic form for g (r):

$$\lim_{r\to\infty} g(r) = \lambda = \left(\frac{|a|}{2b}\right)^{1/2} \tag{5.108}$$

Namely, out of the region of interest the system is in the superconductor state. With (5.108), equation (5.106) requires

$$A(r) \sim \frac{n}{r} \qquad at \qquad r = \infty$$

Thus, we parameterize

$$eA(r) = \frac{n}{r} + ea(r). \tag{5.109}$$

Solving the couple differential equations can only be done numerically. Here, we obtain only the asymptotic form, i.e. at $r \approx \infty$ and $r \approx 0$.

For the case of type II, especially in the case of $m_V \gg m_S$ we expect that the asymptotic value of (5.107) is reached much before the magnetic field since we expect the appreciable functional change in $g(r)$ is at $1/m_S$ and that of $a(r)$ at $1/m_V$. So by setting $g \approx \lambda$ we obtain the following equation for $a(r)$:

$$a(r) = cK_1(e\lambda r) \underset{r \to \infty}{\approx} \left(\frac{\pi}{2e\lambda r} \right)^{1/2} e^{-e\lambda r} \tag{5.110}$$

where K_1 is the first modified Bessel function. Using (5.105) we obtain the following asymptotic form of magnetic field

$$H(r) \approx \frac{c}{e} \left(\frac{\pi\lambda}{2er} \right)^{1/2} e^{-m_V r / 2^{1/2}} \tag{5.111}$$

The asymptotic form of $g(r)$ is also obtained in a similar manner by solving (5.106) asymptotically:

$$g(r) \underset{r \to \infty}{\approx} \lambda + e^{-m_S r / 2^{1/2}} \tag{5.112}$$

Near $r \sim 0$, ϕ must be regular. Since the phase of ϕ is given by $2\pi n\,\theta$ (see (5.104)), $g(r)$ must approaches to zero as

$$g(r) \sim r^n \tag{5.113}$$

Thus, the vortex solution looks as in Fig. 5-3. The region (I) is the superconducting phase and there $H \sim 0$ and $|\phi| \sim \lambda$, while in region (II) a vortex is formed and $H \neq 0$.

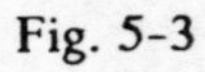
Fig. 5-3

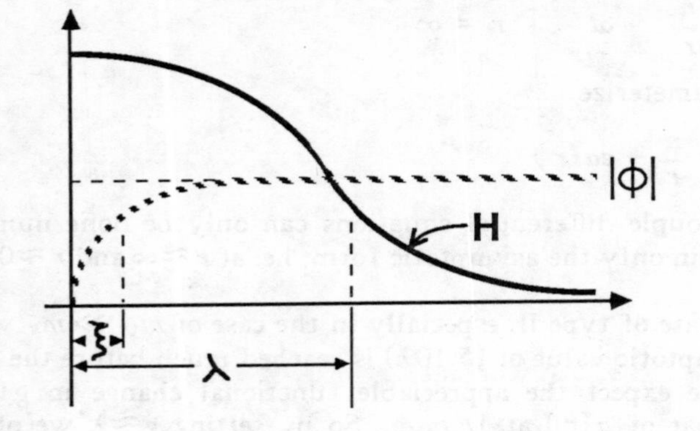
|Φ|
H
ξ
λ

VI. Point Canonical Transformation

In physics we encounter problems which can be handled much better after making an appropriate change of variables. In field theory also the change of variables plays an important role. In this lecture we formulate a method of point canonical transformation which will be useful for the non-perturbative calculations discussed in later lectures.

6.1 Point Canonical Transformation in Operator Formalism:

We start with a standard form of Hamiltonian (see (1.11) here we use indices $a, b, \cdots$ instead of $m, n, \cdots$). In the coordinate representation, the Schrödinger equation is given by

$$\hat{H}\psi = \left[-\frac{1}{2}\sum_{a=1}^{N} \frac{\partial^2}{\partial q^{a2}} + V(\vec{q}) \right] \psi(q) = E\psi(q) \tag{6.1}$$

Let us consider a point canonical transformation given by

$$q^a \to Q^a = f^a(q) \tag{6.2}$$

We assume that the inverse exists:

$$q^a = F^a(Q) \tag{6.3}$$

We express Schrödinger equation (6.1) in terms of Q^a and its derivative. We first insert $F^a(Q)$ in the place of q^a in the wave function and regard it as a function of Q. Then we use the chain rule of differentiation to convert the derivatives.

$$\frac{\partial}{\partial q^a}\psi(q) = \sum_b \frac{\partial f^b(q)}{\partial q^a} \frac{\partial}{\partial Q^b}\psi(F(Q)) \tag{6.4}$$

$$-\frac{1}{2}\sum_a \frac{\partial^2}{\partial q^{a2}}\psi(q) = \frac{1}{2}\left[-i\sum_{ab}\frac{\partial^2 f^b}{\partial q^{a2}} \frac{\partial}{i\partial Q^b} \right.$$

$$\left. + \sum_{abc} \frac{\partial f^b}{\partial q^a} \frac{\partial f^c}{\partial q^a} \frac{\partial}{i\partial Q^b} \frac{\partial}{i\partial Q^c} \right] \psi(F(Q)) \tag{6.5}$$

We define

$$\omega^a(Q) \equiv -\sum_b \frac{\partial^2 Q^a}{\partial q^{b2}} = -\sum_b \frac{\partial^2 f^a}{\partial q^{b2}} \tag{6.6}$$

and

$$\Omega^{ab}(Q) \equiv \sum_c \frac{\partial Q^a}{\partial q^c}\frac{\partial Q^b}{\partial q^c} = \sum_c \frac{\partial f^a}{\partial q^c}\frac{\partial f^b}{\partial q^c} \tag{6.7}$$

Then we obtain

$$\hat{H}\psi = \left|\frac{1}{2}(i\sum_a \omega^a(Q)P_a + \sum_{ab}\Omega^{ab}(Q)P_a P_b) + \tilde{V}(Q)\right|\psi(F(Q)) = E\psi(F(Q)) \tag{6.8}$$

where we set

$$\frac{1}{i}\frac{\partial}{\partial Q^a} = P_a \tag{6.9}$$

$$\tilde{V}(Q) \equiv V(F(Q)) \tag{6.10}$$

The Hamiltonian after the change of variables appears to be non-Hermitian if we take the following naive Hermitian conjugate:

$$P_a^+ = P_a, \qquad Q^{a+} = Q^a \tag{6.11}$$

This is because H is Hermitian in the original q-space. But after the change of variables, the Q-space should be defined by multiplying the wave function by the square root of the Jacobian in order to satisfy the naive Hermitian conjugate (6.11):

$$\int dq\, \psi_1^*(q)\psi_2(q) = \int J(Q)dQ\, \psi_1^*(F(Q))\psi_2(F(Q))$$

$$= \int dQ\, \Psi_1^*(Q)\Psi_2(Q) \tag{6.12}$$

where

$$\Psi(Q) = J^{1/2}(Q)\psi(F(Q)) \tag{6.13}$$

The Hamiltonian in Q-space is then obtained by a similarity transformation

$$H_{eff} = J^{1/2}HJ^{-1/2} \tag{6.14}$$

which should be Hermitian.

In practice the Jacobian is difficult to calculate while ω and Ω

defined by (6.6) and (6.7) are relatively easy. So, it would be nice if H_{eff} is expressed in terms of ω and Ω. Note first

$$J^{+}(Q) = J(Q^{+}) = J(Q) \tag{6.15}$$

Thus,

$$J^{1/2} P_a J^{-1/2} = P_a + iC_a(Q) \tag{6.16}$$

$$C_a(Q) = \frac{1}{2}\frac{\partial}{\partial Q^a}\ln J(Q), \qquad C_a^{+} = C_a \tag{6.17}$$

$$H_{eff} = \frac{1}{2}\left(i\sum_a \omega^a(Q)(P_a + iC_a) + \sum_{ab}\Omega^{ab}(P_a + iC_a)(P_b + iC_b)\right) + \tilde{V}(Q) \tag{6.18}$$

Since H_{eff} should be Hermitian,

$$H_{eff} - H_{eff}^{+} = i\sum_a(\omega^a + 2\sum_b \Omega^{ab} C_b + \sum_b \Omega^{ab}_{,b})(P_a - \frac{1}{2} iC_a) = 0 \tag{6.19}$$

we obtain

$$\omega^a + 2\sum_b \Omega^{ab} C_b + \sum_b \Omega^{ab}_{,b} = 0 \tag{6.20}$$

This is the equation to determine C_a . H_{eff} is then computed as

$$H_{eff} = \frac{1}{2}\sum_{ab}(P_a\,\Omega^{ab} P_b + C_a\,\Omega^{ab} C_b) + \tilde{V} \tag{6.21}$$

(6.20) and (6.21) are the main results, which will be used in the later lectures.

6.2 Weyl Ordering and Midpoint Prescription of Path Integral

The Hamiltonian (6.21) contains P's and Q's in a product form. (Note Ω is a function of Q's.) However, since we obtained it from the standard form of Hamiltonian by a point canonical transformation via chain rule of differentiation, the operator ordering is fixed as in (6.21). Next we consider a path integral expression of Feynman kernel associated with this Hamiltonian. For this purpose let us first define a specific operator ordering, the Weyl ordering.

Weyl Ordering:

Weyl ordering is defined in the following way

$$(\hat{p}\hat{q})_W = \frac{1}{2}(\hat{p}\hat{q} + \hat{q}\hat{p})$$

$$(\hat{p}\hat{q}^3)_W = \frac{1}{4}(\hat{p}\hat{q}^3 + \hat{q}\hat{p}\hat{q}^2 + \hat{q}^2\hat{p}\hat{q} + \hat{q}^3\hat{p})$$

etc.

General expression =

$$= \frac{\sum(\textit{all possible orders})}{\textit{total number of possible orders}} \tag{6.22}$$

The following expression is a generating function:

$$(\alpha\hat{p} + \beta\hat{q})^N = \sum_{\substack{l,m \\ l+m=N}} \frac{N!}{l!m!}\alpha^l \beta^m (\hat{p}^l \hat{q}^m)_W \tag{6.23}$$

Midpoint Prescription and Weyl Ordering:

Next we prove

$$\langle q \mid H(\hat{p},\hat{q})_W \mid q'\rangle = \int \frac{dp}{2\pi} e^{ip(q-q')} H(p, \frac{q+q'}{2}) \tag{6.24}$$

For this it is sufficient to prove

$$\langle q \mid (\alpha\hat{p} + \beta\hat{q})^N \mid q'\rangle = \int \frac{dp}{2\pi} e^{ip(q-q')} (\alpha p + \beta \frac{q+q'}{2})^N \tag{6.25}$$

The proof we use is by mathematical induction. For N = 1,

$$\langle q \mid (\alpha\hat{p} + \beta\hat{q}) \mid q'\rangle = \int \frac{dp}{2\pi} \langle q \mid (\alpha\hat{p} + \beta\hat{q}) \mid p\rangle\langle p \mid q'\rangle$$

$$= \int \frac{dp}{2\pi} e^{ip(q-q')}(\alpha p + \beta q)$$

Since it is obvious βq can be replaced by $\beta\frac{q+q'}{2}$, (6.25) holds for N = 1. Next, assuming (6.25) for N = n, we compute it for N= n+1.

$$<q\,|(\alpha\hat{p} + \beta\hat{q})^{n+1}|\, q'> = (\frac{\alpha}{i}\frac{\partial}{\partial q} + \beta q)\, <q\,|(\alpha\hat{p} + \beta\hat{q})^{n}\, |q'>$$

$$= (\frac{\alpha}{i}\frac{\partial}{\partial q} + \beta q)\int \frac{dp}{2\pi} e^{ip(q-q')}(\alpha p + \beta\frac{q+q'}{2})^{n} \ (assumption)$$

$$= \int \frac{dp}{2\pi} e^{ip(q-q')}(\alpha p + \beta\frac{q+q'}{2} + \frac{\alpha}{i}\frac{\partial}{\partial q} + \beta\frac{q-q'}{2})\,(\alpha p + \beta\frac{q+q'}{2})^{n}$$

The last term in the integral can be replaced by $-\frac{\beta}{2i}\frac{\partial}{\partial p}$ by using the integration by part. So, for N = n + 1 (6.25) holds since

$$\frac{1}{i}(\alpha\frac{\partial}{\partial q} - \frac{\beta}{2}\frac{\partial}{\partial p})(\alpha p + \beta\frac{q+q'}{2})^{n} = 0$$

(QED)

At this point let us go back to **2.1** and repeat the path integral conversion procedure. Then we convince ourselves that the midpoint prescription of path integral expression (2.12) is the phase space path integral representation of Feynman kernel of Weyl ordered Hamiltonian. Therefore, in order to write the path integral representation for the Feynman kernel of effective Hamiltonian (6.21), we must Weyl order the Hamiltonian first. Since

$$P_a\,\Omega^{ab}\,P_b = (P_a\,\Omega^{ab}\,P_b)_W + \frac{1}{4}\Omega^{ab}_{,ab} \tag{6.26}$$

we obtain

$$H_{eff} = \frac{1}{2}\sum_{ab}(P_a\,\Omega^{ab}\,P_b)_W + \tilde{V}(Q) + \Delta V(Q) \tag{6.27}$$

where

$$\Delta V = \sum_{ab}\frac{1}{2}\,(\,C_a\,\Omega^{ab}\,C_b + \frac{1}{4}\Omega^{ab}_{,ab}) \tag{6.28}$$

ΔV is an extra term which does not appear in the classical point canonical transformation. It is a quantum effect and if we keep $\hbar$ in the calculation we see that ΔV is proportional to $\hbar^2$.

Now the path integral expression of Feynman kernel is given by

$$<Q_f \mid e^{-i\hat{H}_{eff}(t_f - t_i)} \mid Q_i> =$$

$$= \lim_{N\to\infty} \int_{\substack{Q^a(N)=Q^a_f \\ Q^a(0)=Q^a_i}} \cdots \int \left[\prod_{a=1}^{M} \prod_{n=1}^{N-1} dQ(n) \right] \left[\prod_{n=1}^{N} \frac{dP^a(n)}{2\pi} \right] e^{i\sum_n A_n} \tag{6.29}$$

$$A_n = \sum_a P_a(n)(Q^a(n) - Q^a(n-1)) - \epsilon H_n$$

$$H_n = \frac{1}{2}\sum_{a,b} P_a(n)P_b(n)\Omega^{ab}(\bar{Q}(n)) + V(\bar{Q}(n)) + \Delta V(\bar{Q}(n)) \tag{6.30}$$

where $\bar{Q}(n)$ is given by

$$\bar{Q}(n) = \frac{1}{2}(\bar{Q}(n) + \bar{Q}(n+1)) \tag{6.31}$$

In order to obtain the path integral expression in configuration space we integrate over P, which is a Gaussian integration. We obtain

$$<Q_f \mid e^{-i\hat{H}_{eff}(t_f - t_i)} \mid Q_i >$$

$$= \lim_{N\to\infty} \int_{\substack{Q(0)=Q_i \\ Q(N)=Q_f}} \cdots \int \left[\prod_{a=1}^{M} \prod_{n=1}^{N-1} dQ^a(n) \right] (2\pi i \epsilon)^{-NM/2} \left[\prod_{n=1}^{N} \det \Omega(\bar{Q}(n)) \right]^{-1/2} e^{i\epsilon\sum_n L_n} \tag{6.32}$$

where

$$L_n = \frac{1}{2\epsilon^2}\sum_{ab} \left[Q^a(n) - Q^a(n-1) \right] \Omega^{-1}_{ab}(\bar{Q}(n)) \left[Q^b(n) - Q^b(n-1) \right]$$

$$- \tilde{V}(\bar{Q}(n)) - \Delta V(\bar{Q}(n)) \tag{6.33}$$

In short hand notation

$$<Q_f \mid e^{-iH_{eff}(t_f - t_i)} \mid Q_i >$$

$$= \int \cdots \int DQ(\det\Omega)^{-1/2} e^{i\int_{t_i}^{t_f} L(\dot{Q},Q)dt} \tag{6.34}$$

Notice the appearance of $(\det\Omega)^{-1/2}$ in the path integration measure.

ΔV :

The point canonical transformation (6.2) can be viewed as a transformation from Cartesian coordinates to curved coordinates. Since

$$(ds)^2 = \sum_a (dq^a)^2 = \sum_{ab} dQ^a dQ^b \sum_c \frac{\partial q^c}{\partial Q^a} \frac{\partial q^c}{\partial Q^b} \equiv \sum_{ab} dQ^a dQ^b g_{ab} \quad (6.35)$$

the metric in Q-system is given by

$$g_{ab} \equiv \Omega_{ab}^{-1}, \qquad g^{ab} = \Omega_{ab} \quad (6.36)$$

(see (6.7)). Thus,

$$J^2 = \det g \equiv g, \qquad C_a = \frac{1}{4}(\ln g)_{,a} \qquad ,etc. \quad (6.37)$$

From now on we use the standard summation convention for the repeated indices. After some calculations (6.28) is expressed entirely in terms of g:

$$\Delta V = \frac{1}{8} \Gamma_{ac}^{b} \Gamma_{bd}^{a} g^{cd} \quad (6.38)$$

where Γ is a Riemann Christoffel symbol of second kind:

$$\Gamma_{bc}^{a} = \frac{1}{2} g^{ad} (g_{bd,c} + g_{dc,b} - g_{bc,d}) \quad (6.39)$$

6.3 Point Canonical Transformation in Path Integral

In **6.1** we studied the point canonical transformation in operator formalism, specifically the change of variables of a standard form of Hamiltonian. Then we converted the Feynman kernel of the resulting effective Hamiltonian into a path integral expression (6.32). In this section we derive the same result starting from the path integral expression (2.15) by extending it to M variables first and then changing variables within the path integral. The method we use is due to Gervais and Jevicki.

The Feynman kernel to go from some initial configuration $q_i = (q_i^1, q_i^2, \ldots, q_i^M)$ at time t_i to q_f at t_f is given by

$$<q_f \mid e^{-i\hat{H}(t_f - t_i)} \mid q_i>$$

$$= \lim_{N \to \infty} \int_{\substack{q(t_f)=q_f \\ q(t_i)=q_i}} \cdots \int \left[\frac{\prod_{a=1}^{M} \prod_{n=1}^{N-1} dq_n^a}{(2\pi i \epsilon)^{\frac{NM}{2}}} \right] e^{i\epsilon \sum_n L_n} \tag{6.40}$$

where

$$L_n = \frac{1}{2\epsilon^2} \sum_a \left[q^a(n) - q^a(n-1) \right]^2 - V(\bar{q}(n)) \tag{6.41}$$

We then change variables: $q \rightarrow Q$

$$q^a(n) = F^a(Q(n)) \tag{6.42}$$

The new integration measure is given by

$$\prod_{a=1}^{M} dq^a(n) = g^{1/2}(Q(n)) \prod_{a=1}^{M} dQ^a(n) \tag{6.43}$$

The Lagrangian is then

$$L_n = \frac{1}{2\epsilon^2} \left[F^a(Q(n)) - F^a(Q(n-1)) \right]^2 - V\left[F(Q(n)) \right] \tag{6.44}$$

We expand (6.44) about the mid-point $\bar{Q}(n)$ given by (6.31). The leading kinetic energy term is obviously given by

$$L_n^{(0)} = \frac{1}{2\epsilon^2} g_{ab}(\bar{Q}(n)) \, \Delta Q^a(n) \, \Delta Q^b(n) \tag{6.45}$$

where

$$\Delta Q^a(n) = Q^a(n) - Q^a(n-1) \tag{6.46}$$

In the expansion of (6.44) up to which power of ΔQ should be kept is determined by the estimate of ΔQ. Since the path integral becomes eventually a Gaussian integration the estimate of ΔQ can be done by estimating the average of $(\Delta Q)^2$ with the action given by (6.45):

$$< (\Delta Q)^2 > \propto \int d(\Delta Q) \, (\Delta Q)^2 \, e^{i\epsilon \frac{1}{2\epsilon^2} g (\Delta Q)^2} \approx O(\epsilon) \tag{6.47}$$

Since all the terms up to ϵ^0 order in L_n contribute to the action, we must keep up to quartic terms of ΔQ in the expansion of (6.44). By a straightforward calculation we obtain

$$L_n \approx \frac{1}{2\epsilon^2} g_{ab}(\bar{Q}(n)) \Delta Q^a \Delta Q^b +$$

$$+ \frac{1}{24\epsilon^2} F^a_{,b}(\bar{Q}(n))\, F^a_{,cde}\,(\bar{Q}(n))\Delta Q^b \Delta Q^c \Delta Q^d \Delta Q^e \; -$$

$$- \tilde{V}(\bar{Q}(n)) \qquad (6.48)$$

where $\quad \tilde{V}(Q) = V(F(Q)) \quad$ (see (6.10))

Next we expand the Jacobian about the mid-point. Since the total Jacobian is given by

$$\prod_{n=1}^{N-1} g^{1/2}(Q(n))) = g^{-1/4}(Q_f\,)g^{-1/4}(Q_i\,) \prod_{n=1}^{N} g^{1/4}(Q(n))\, g^{1/4}(Q(n-1)) \qquad (6.49)$$

We expand $\left[g(Q(n))\, g(Q(n-1))\right]^{1/4}$ by using

$$\det(A+B) = \det A \;\det(1 + A^{-1}B) =$$

$$= \det A \;[\, 1 + tr(A^{-1}B) + \frac{1}{2}(trA^{-1}B)^2 - \frac{1}{2}tr(A^{-1}B)^2 + \ldots.\,] \qquad (6.50)$$

We obtain after some calculation

$$g^{1/4}(Q(n))\, g^{1/4}(Q(n-1)) \approx$$

$$\approx g^{1/2}(\bar{Q}(n)) \left[1 + \frac{1}{16}\left[g^{ab}(\bar{Q}(n)) g_{ab,cd}(\bar{Q}(n)) + g^{ab}_{,c}(\bar{Q}(n))\, g_{ab,d} \right] \Delta Q^c(n) \Delta Q^d(n) \right] \qquad (6.51)$$

Substituting (6.43), (6.48), (6.49) and (6.51) into (6.40) and rearranging the factors we obtain the following expression as an effective Jacobian.

$$g^{1/2}(\bar{Q})[\, 1 + \frac{1}{16}\,(g^{ab}(\bar{Q})\, g_{ab,cd}(\bar{Q}) + g^{ab}_{,c}(\bar{Q})\, g_{ab,d}(\bar{Q}))\, \Delta Q^c \Delta Q^d\, +$$

$$+ \frac{i}{24\epsilon} F^e_{,a} F^e_{,bcd}(\bar{Q})\, \Delta Q^a \Delta Q^b \Delta Q^c \Delta Q^d \;] \qquad (6.52)$$

The last term of (6.52) is the contribution due to the expansion of L_n (6.48). We replace ΔQ's by the average with the weight $e^{i\epsilon \sum_n L_n^{(0)}}$. The justification of this procedure is not very simple but goes as follows. Let us focus our attention on the integration of $Q(n)$ and $Q(n-1)$. We change variables to $\bar{Q}(n)$ and $\Delta Q(n)$. Then perform ΔQ integration which is equivalent to the average. There should be a contribution due to $\bar{Q}(n\pm 1)$ and $\Delta Q(n\pm 1)$, since they also depend on $Q(n)$ and $Q(n-1)$. However, this contribution is shown to be smaller by an amount of order ϵ and therefore negligible. We use

$$<\Delta Q^a \Delta Q^b> = i\epsilon g^{ab}$$

$$<\Delta Q^a \Delta Q^b \Delta Q^c \Delta Q^d>$$

$$= (i\epsilon)^2 (g^{ab} g^{cd} + g^{ac} g^{bd} + g^{ad} g^{bc}) \tag{6.53}$$

for (6.52) and exponentiate it to have an additional term in the action. We obtain

$$\Delta V(\bar{Q}) = -\frac{1}{16}(g^{ab} g_{ab,cd} + g^{ab}_{,c} g_{ab,d}) g^{cd} +$$

$$+\frac{1}{24} F^e_{,a} F^e_{,bcd}(g^{ab} g^{cd} + g^{ac} g^{bd} + g^{ad} g^{bc}) \tag{6.54}$$

The following expressions are used to prove that (6.54) is the same as (6.38):

$$g^{-1} g_{,c} = g^{ab} g_{ab,c}$$

$$F^a_{,cd} = \Gamma^e_{cd} F^a_{,e}$$

$$F^e_{,a} F^e_{,bcd} g^{ab} g^{cd} = \Gamma^a_{ab,c} g^{bc} + g^{ab} \Gamma^c_{da} \Gamma^d_{cb}$$

$$\Gamma^a_{ab,c} = \frac{1}{2}(g^{-1} g_{,b})_{,c} \tag{6.55}$$

The final form of the path integral expression is identical to (6.32) except the factor $g^{-1/4}(Q_f) g^{-1/4}(Q_i)$ which should be the Jacobian factor between $|Q>$ and $|q>$. (See (6.13). Note $J = g^{1/2}$)

Exercise: Work out all the calculations left out in this section.

6.4 Perturbation Expansion in Phase Space Path Integral

We discussed the perturbation expansion in **IV**. The discussion used there can be used for the Feynman kernel such as (6.34) discussed in the previous section. Once this is done we obtain an unambiguous result. Thus the perturbation description must contain the operator ordering information somewhere. One of the reasons to add this section is to show that this information is contained in the propagator used in the perturbation expansion.

The other reason is a more practical one, namely to develop a perturbation expansion technique in phase space path integral. This should be useful especially for path integrals such as (6.29), because the corresponding configuration space path integral (6.32) involves a determinant in the intergrand so that the standard Feynman prescription should contain a set of vertices (usually an infinite number) due to $\ln \det \Omega$ (by exponentiating the determinant) and becomes rather complicated.

Generating functional is expressed as

$$Z[J,K] = \int Dq \int \frac{Dp}{2\pi} e^{i[A(p,q) + \int_{-\infty}^{\infty} dt [J(t)q(t) + K(t)p(t)]} \tag{6.56}$$

Source functions $J(t)$ and $K(t)$ are arbitrary functions of t, with a restriction that they vanish at $t = \pm\infty$. Let us discuss one-variable problem first. Then, the action is given by

$$A(p,q) = \int_{-\infty}^{\infty} dt [p(t)\dot{q}(t) - H(p(t),q(t))] \tag{6.57}$$

In order to develop perturbation theory we assume

$$H = H_0 + H_1$$

$$H_0 = \frac{1}{2}(p^2 + \omega^2 q^2)$$

$$H_1 = H_1(p,q) \tag{6.58}$$

As before (see (4.12)) we obtain

$$Z[J,K] = e^{i\int_{-\infty}^{\infty} dt H_1[\frac{1}{i}\frac{\partial}{\partial K}, \frac{1}{i}\frac{\partial}{\partial J}]} Z_0[K,J] \tag{6.59}$$

where Z_0 is given by

$$Z_0 = \int Dq \int \frac{Dp}{2\pi} e^{i\int_{-\infty}^{\infty} dt [p\dot{q} - \frac{1}{2}(p^2 + \omega^2 q^2) + Jq + Kp]} \tag{6.60}$$

We make a shift,

$$q \to q + \Delta J , \qquad p \to p + \dot{\Delta K} \tag{6.61}$$

and bring Z_0 to the following form:

$$\exp[-\frac{1}{2}\int_{-\infty}^{\infty} dt \int_{-\infty}^{\infty} dt'[(J(t) - \dot{K}(t))\Delta_F(t-t')(J(t') - \dot{K}(t'))] + \frac{i}{2}\int_{-\infty}^{\infty} dtK^2(t)] \tag{6.62}$$

assuming the Feynman boundary condition.

$$\Delta_F = i\int_{-\infty}^{\infty}\frac{d\nu}{2\pi}\frac{e^{-i\nu t}}{\nu^2-\omega^2+i\epsilon} = \frac{1}{2\omega}[\theta(t)e^{-i\omega t} + \theta(-t)e^{i\omega t}] \tag{6.63}$$

is Feynman's Green function. Thus, the propagators are

$$\overline{q(t)q}(t') = \frac{1}{i^2}\frac{\delta^2 Z_0}{\delta J(t)\delta J(t')} = \Delta_F(t-t')$$

$$\overline{q(t)p}(t') = \frac{1}{i^2}\frac{\delta^2 Z_0}{\delta J(t)\delta K(t')} = \frac{\partial}{\partial t'}\Delta_F(t-t')$$

$$\overline{p(t)p}(t') = \frac{1}{i^2}\frac{\delta^2 Z_0}{\delta K(t)\delta K(t')} = \frac{\partial}{\partial t}\frac{\partial}{\partial t'}\Delta_F(t-t') - i\delta(t-t') \tag{6.64}$$

which we denote by the lines shown in Fig.6-1.

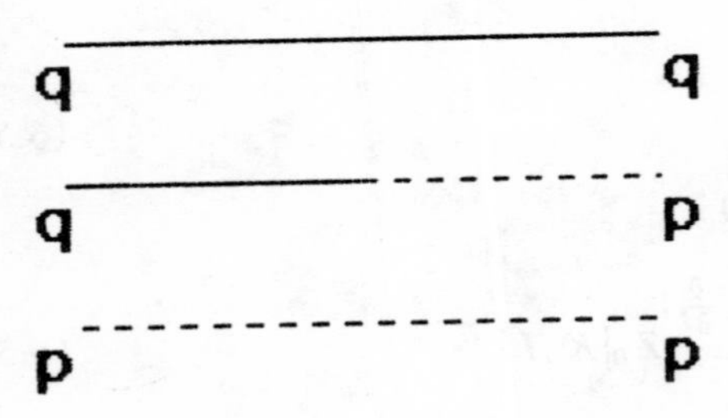

Fig.6-1

We note first

$$\frac{\partial}{\partial t}\Delta_F(t)\Big|_{t=0} = \int_{-\infty}^{\infty} \frac{id\nu}{2\pi}\,\frac{-i\nu}{\nu^2-\omega^2+i\epsilon} = 0 \tag{6.65}$$

so that $q(t)p(t)=0$. This result obviously indicates that our Feynman rules are associated with a specific operator ordered Hamiltonian. In fact, it is a Weyl ordered Hamiltonian.

To prove this we must use discrete treatment and take the limit afterwards

$$Z[J,K] = \lim_{\substack{t_f\to\infty \\ t_i\to-\infty}} \int\int dq_f\, dq_i\, \psi_0^*(q_f,t_f)K(q_f,t_f,q_i,t_i)\psi_0(q_i,t_i) \tag{6.65}$$

where K is the Feynman kernel, ψ_0 the asymptotic ground state wave function of the free Hamiltonian, $H_0 = \frac{1}{2}(p^2+\omega^2q^2)$:

$$K = \int\cdots\int \prod_{n=1}^{N-1} dq(n) \prod_{n=1}^{N} \frac{dp(n)}{2\pi}\, e^{i\sum_n [A_n + \epsilon J(n)q(n) + \epsilon K(n)p(n)]} \tag{6.67}$$

$$A_n = \epsilon(p(n)\dot q(n) - H_n) \tag{6.68}$$

$$H_n = H_0(n) + H_1(p(n), \frac{q(n)+q(n-1)}{2}) \tag{6.69}$$

$$\epsilon = \frac{1}{N}(t_f - t_i) \tag{6.70}$$

$$\psi_0(q,t) = (\frac{\omega}{\pi})^{1/4} e^{-\frac{1}{2}\omega q^2 - \frac{i}{2}\omega t} \tag{6.71}$$

The calculation of $Z_0[J,K]$ is tedious but straightforward. The result is

$$Z_0[J,K] = \lim_{\substack{t_f\to\infty \\ t_i\to-\infty}} (\frac{\eta}{2})^{\frac{N}{2}} \left[ie^{i\omega(t_f-t_i) - N\alpha}\right]^{1/2} \times$$

$$\exp\left[-\frac{\epsilon^2}{2}\sum_{n,m=1}^{N}[J(n)-\dot{K}(n+1)]\Delta_F(n,m)[J(m)-\dot{K}(m+1)]+\frac{i}{2}\epsilon\sum_{n=1}^{N}K^2(n)\right] \tag{6.72}$$

where

$$\eta=\frac{8\epsilon}{4+\omega^2t^2}=\frac{2}{\omega}\sin\alpha \qquad \tan\frac{\alpha}{2}=\frac{\epsilon\omega}{2} \tag{6.73}$$

and

$$\Delta_F(n,m)=\frac{1}{2\omega}[\theta(n-m-1)e^{-i(n-m)\alpha}+\delta_{n,m}+\theta(m-n-1)e^{-i(m-n)\alpha}] \tag{6.74}$$

is the discrete version of Feynman propagator with

$$\theta_k = \begin{cases} 1 & n\geqslant 0 \\ 0 & n\leqslant -1 \end{cases} \tag{6.75}$$

So in continuum limit

$$\overline{q(n)\,q(m)} \underset{\epsilon\to 0}{\to} \begin{cases} \frac{1}{2\omega}e^{-i\omega(t-t')} & n\geqslant m \qquad t\geqslant t' \\ -\frac{1}{2\omega}e^{-i\omega(t'-t)} & n\leqslant m-1 \quad t\leqslant t' \end{cases} \tag{6.76}$$

where

$$t_i+m\epsilon=t$$

$$t_i+n\epsilon=t' \tag{6.77}$$

Thus we obtained precisely the Feynman propagator

$$\Delta_F=\frac{1}{2\omega}[\theta(t)\,e^{-i\omega t}+\theta(-t)e^{i\omega t}] \tag{6.78}$$

Now, the crucial point is

$$\overline{q(n)\,p(m)} \underset{\epsilon\to 0}{\to} \begin{cases} \frac{i}{2}e^{-i\omega(t-t')} & n\geqslant m \qquad t\geqslant t' \\ -\frac{i}{2}e^{-i\omega(t'-t)} & n\leqslant m-1 \quad t\leqslant t' \end{cases} \tag{6.79}$$

Thus,

$$\overline{q(n)p(n)} = \frac{i}{2}$$

$$\overline{q(n-1)p(n)} = -\frac{i}{2}$$

and adding these two expressions we obtain the desired result

$$\overline{\frac{q(k)+q(k-1)}{2}p_k} = 0 \tag{6.80}$$

From (6.65) and (6.80) we conclude that the Feynman rules given by (6.64) are valid only when the path integral is defined in terms of the mid-point prescription. As we discussed in the previous section, the Hamiltonian used in the phase space path integral with mid-point prescription is a Weyl ordered one. Therefore, for the phase space perturbation scheme described in this section one must use a Weyl ordered Hamiltonian.

VII. Large N Collective Variables

In this chapter we apply the method of change of variables developed in the previous chapter to the large N problems. The essence of the method is the change of original dynamical variables to a set of so-called "collective variables". The collective variables are the invariant combinations of the original variables of the system which we assume has a symmetry and the low energy states are singlets of the symmetry. The method is a natural extension of the Bohm-Pines theory of high density plasma oscillations. So we begin with the high density collective oscillations of Bose particles.

7.1. The Collective Field Theory of N Bose Particles

To illustrate in detail the main ideas, let us consider a quantum mechanical system of N Bose particles in one dimension of size L with periodic boundary conditions. The Hamiltonian representing the system is taken to be

$$H = \frac{1}{2}\sum_{i=1}^{N}\hat{p}_i^2 + \frac{1}{2}\sum_{i\neq j}^{N} v(\hat{x}^i, \hat{x}^j) + \sum_{i=1}^{N} V(\hat{x}^i) \tag{7.1}$$

The second term is the two particle interaction energy, while the third is the energy due to a common potential $V(x)$.

The Schrödinger wave function is a totally symmetric function of x_j $(j = 1, \cdots, N)$ (Bose symmetry). Since the density operators

$$\hat{\phi}(x) = \frac{1}{N}\sum_{i=1}^{N}\delta(x - \hat{x}^i), \qquad \frac{1}{2}L \geqslant x \geqslant -\frac{1}{2}L \tag{7.2}$$

are the most general commuting symmetric operators, one may regard, in general, the wave function as a functional of the density function $\phi(x)$;

$$\Psi(x_1, x_2, \ldots, x_n) = \Phi[\phi(\cdot)] \tag{7.3}$$

Notice that equation (7.3) restricts to totally symmetric wave functions so that the Bose statistics is already imposed.

The number of degrees of freedom of the system is N. On the other hand the number of $\hat{\phi}(x)$ with continuous x is infinite. This merely means that all $\phi(x)$ are not necessarily independent. Especially, it is obvious that

$$\int dx\, \hat{\phi}(x) = 1 \tag{7.4}$$

Thus, in order to choose a set of independent variables we should use Fourier components:

$$\hat{\phi}_k = \int dx\, e^{-ikx}\, \hat{\phi}(x) = \frac{1}{N}\sum_{i=1}^{N} e^{-ik\hat{x}^i} \tag{7.5}$$

where k takes discrete values

$$k = \frac{2\pi n}{L}, \qquad n = \pm 1, \pm 2, \cdots \tag{7.6}$$

We restrict

$$|k| \leqslant k_{\max} = \frac{\pi N}{L} \tag{7.7}$$

and choose $\hat{\phi}_k$ as a set of new variables (collective variables).

Since N/L is the density of Bose particles, the high density limit is $k_{\max} \to \infty$. Therefore, in this limit one regards $\hat{\phi}$ as independent variables with constraint (7.4). From the definition of (7.2) we see also that the eigenvalues of $\hat{\phi}$ are restricted to be positive:

$$\phi(x) \geqslant 0 \qquad \frac{1}{2}L \geqslant x \geqslant -\frac{1}{2}L \tag{7.8}$$

Next we change variables from x's to ϕ_k's following the discussion of the previous chapter. We first calculate ω and Ω defined by (6.6) and (6.7).

$$\omega(k;[\phi]) = -\sum_{i=1}^{N}\frac{\partial^2}{\partial x^{i2}}\frac{1}{N}\sum_{i=1}^{N} e^{-ikx^i} = k^2\phi_k \tag{7.9}$$

$$\Omega(k,-k';[\phi]) = \sum_{i=1}^{N}\left(\frac{\partial}{\partial x^i}\phi_k\right)\left(\frac{\partial}{\partial x^i}\phi_{-k'}\right) = \frac{kk'}{N}\phi_{k-k'} \tag{7.10}$$

Then we substitute them in (6.20) to obtain C^a. In the present case

$$\frac{\partial\Omega(k,-k',[\phi])}{\partial\phi_k} = \frac{kk'}{N}\,\delta_{k,k-k'} = 0 \tag{7.11}$$

so that the second term of effective Hamiltonian (6.21) is given by

$$V_{coll} = \frac{1}{2}\sum C\,\Omega\,C = \frac{1}{8}\sum_{kk'}\omega(-k;[\phi])\,\Omega^{-1}(k,-k';[\phi])\,\omega(k';[\phi]) \tag{7.12}$$

where Ω^{-1} is the inverse matrix of Ω given by (7.10). The inverse matrix is well defined since the diagonal components of Ω are c-numbers (note $\phi_0 = 1$).

V_{coll} given by (7.12) can be interpreted as the centrifugal barrier energy associated with the Bose symmetry. If we keep $\hbar$ throughout the calculation we notice that V_{coll} is proportional to $\hbar^2$ so that it reflects the quantum nature of this potential. In the high density limit V_{coll} can be expressed as

$$V_{coll} = \frac{N}{8} \int dx \; \frac{(\partial\phi(x))^2}{\phi(x)} \tag{7.13}$$

In this expression $\phi(x)$ appeared in the denominator and looks peculiar. But it is all right since $\phi(x)$ is constrained by (7.4) and (7.8).

In the high density limit the potential term (the second and third term in (7.1) of the Hamiltonian is expressed in terms of collective variables as

$$\tilde{V} = \frac{N^2}{2} \int dx \int dy \; \phi(x)\, v(x,y)\, \phi(y) +$$

$$+ N \int dx\, \phi(x)[V(x) - v(x,x)] \tag{7.14}$$

The kinetic energy term (the first term in (6.21)) is given by

$$K = \frac{1}{2} \sum P \Omega P = \frac{1}{2} \sum_{kk'} \pi_{-k}\, \Omega(k, -k'; [\phi])\, \pi_{k'} \tag{7.15}$$

which in the high density limit becomes

$$K = \frac{1}{2N} \int dx\, (\partial_x \pi)\, \phi(x)\, (\partial_x \pi) \tag{7.16}$$

where π_k is the conjugate momentum to ϕ_{-k} :

$$\pi_k \equiv \frac{1}{i} \frac{\partial}{\partial\phi_{-k}}$$

so that

$$[\pi_k, \phi_{k'}] = -i\delta_{k,-k'} \tag{7.17}$$

Introducing the Fourier transform of π ,

$$\pi(x) = \frac{1}{L} \sum_{k \neq 0} e^{ikx} \pi_k \tag{7.18}$$

we obtain

$$[\pi(x), \phi(x')] = -\sum_{k \neq 0} e^{ik(x-x')}$$

$$\rightarrow -i(\delta(x-x') - \frac{1}{L}) \qquad (\textit{high density limit}) \tag{7.19}$$

The effective Hamiltonian of N Bose system is then given by

$$H_{eff} = K + V_{eff} = K + V_{coll} + \tilde{V} \tag{7.20}$$

where K, V_{coll} and $\tilde{V}$ in the high density limit are given by (7.16), (7.13) and (7.14) respectively.

We apply this formalism to the problem of high density Bose plasma and the collective motions of large number of harmonic oscillators.

High Density Bose Plasma

Let us consider a system of large number of charged Bose particles in the constant negative background charge distribution. We assume the system is neutral so that $V(x)$ is a negative constant times N:

$$V(x) \propto -N \tag{7.21}$$

The two body potential $v(x,x')$ in the present case is the Coulomb potential between a pair of charged Bose particles:

$$v(\vec{x},\vec{x}') = \frac{e^2}{|\vec{x} - \vec{x}'|} \tag{7.22}$$

We can extend the previous formalism to a system in three dimensions without difficulty. We simply replace ∂_x by $\vec{\nabla}$ and L by V (the volume of the system).

In the large N limit $\tilde{V}$ dominates. So we minimize it by making a variation with respect to $\phi(x)$. We obtain

$$N^2 \int v(\vec{x},\vec{x}')\,\phi(\vec{x}')dx' + NV(x') - \lambda = 0 \tag{7.23}$$

where λ is a Lagrange multiplier. Since $V(\vec{x})$ is independent of $\vec{x}$ and since $v(\vec{x},\vec{x}')$ depends only on $|\vec{x} - \vec{x}'|$, it is not difficult to see that the solution of (7.23) which we denote by $\phi^0(\vec{x})$, is also a constant. So, using (7.4) we obtain

$$\phi^0(\vec{x}) = \frac{1}{V} \tag{7.24}$$

Next we expand $\phi(x)$ around (7.24)

$$\phi(x) = \frac{1}{V}[1 + \sum_{\vec{k}\neq 0} (k^2/N)^{1/2} e^{i\vec{k}\cdot\vec{x}} Q_{\vec{k}}] \tag{7.25}$$

$$\pi(x) = \sum_{\vec{k}\neq 0} (N/k^2)^{1/2} e^{-i\vec{k}\cdot\vec{x}} P_{\vec{k}} \tag{7.26}$$

The commutation relation between $Q_{\vec{k}}$ and $P_{\vec{k}}$

$$[P_{\vec{k}}, Q_{\vec{k}'}] = -i\delta_{\vec{k},\vec{k}'} \tag{7.27}$$

reproduces (7.19). The expansion (7.25) is a 1/N expansion so we keep only the quadratic part in H_{eff}. We obtain

$$H_{eff} = \frac{1}{2}\sum_{\vec{k}\neq 0}[\,|P_{\vec{k}}|^2 + \omega^2(k)|Q_{\vec{k}}|^2] \tag{7.28}$$

where

$$\omega^2(k) = \omega_0^2 + \frac{1}{4}(\vec{k}^2)^2 \tag{7.29}$$

$$\omega_0^2 = 4\pi e^2 \rho_0 \equiv 4\pi e^2 N/V \tag{7.30}$$

ρ_0 is the average density of the Bose particles. The effective Hamiltonian (7.28) is a system of free harmonic oscillators of frequency $\omega(k)$. Thus, the excitations of the system are described by the excitations of these oscillators, the plasmons.

The excitation energy is given by (7.29) in which the first term ω_0^2 is due to the Coulomb energy term while the second term $\frac{1}{4}(\vec{k}^2)^2$ is due to the collective potential term V_{coll}. For small $\vec{k}^2$ the plasma frequency is given by ω_0 and it does not involve $\hbar$. Thus this frequency should be derivable from entirely classical considerations. It corresponds to dipole oscillation of a uniform charge distribution against the uniform negative background charge distribution.

Collective Motions of N-Identical Harmonic Oscillators

As an another application of the above formalism let us consider the collective motions of N identical free harmonic oscillators. Obviously this problem allows for an exact solution. However, the physics of the large N limit, namely the collective motions of these oscillators, look quite different.

We have in this case

$$V(x) = \frac{1}{2}\,\omega^2 x^2, \qquad v(x,y) = 0 \tag{7.31}$$

Therefore, the effective potential is now given by

$$V_{eff} = V_{coll} + \tilde{V} = \frac{N}{8}\int dx \frac{(\partial\phi)^2}{\phi(x)} + \frac{N}{2}\,\omega^2 \int x^2 \phi(x)\,dx \tag{7.32}$$

and one must satisfy the constraint condition

$$\int dx\,\phi(x) = 1 \tag{7.33}$$

As before we minimize first V_{eff} . The minimum configuration $\phi^0(x)$ is a solution of

$$\frac{\delta V_{eff}}{\delta\phi(x)} - N\lambda = 0. \tag{7.34}$$

The equation is

$$-\frac{1}{8}\,\frac{(\partial\phi)^2}{\phi^2} - \frac{1}{4}\,\partial\left(\frac{\partial\phi}{\phi}\right) + \frac{\omega^2}{2}x^2 - \lambda = 0 \tag{7.35}$$

It is not difficult to see that ϕ^0 is given by

$$\phi^0(x) = \left(\frac{\omega}{\pi}\right)^{1/2} e^{-\omega x^2} \tag{7.36}$$

Next we expand $\phi(x)$ around ϕ^0 by setting

$$\phi(x) = \phi^0(x) + \frac{1}{\sqrt{N}}\,\eta(x) \tag{7.37}$$

$$\pi(x) = \sqrt{N}\,\xi(x) \tag{7.38}$$

The scale factors $\sqrt{N}$ and $\frac{1}{\sqrt{N}}$ in (7.37) and (7.38) are determined such that after the expansion of H_{eff} the quadratic parts of K and V_{eff} are the same order in N and $\eta(x)$ and $\xi(x)$ satisfy the same canonical commutation relation:

$$[\,\xi(x),\, \eta(x')] = -i\,(\delta(x - x') + const.) \tag{7.39}$$

After some calculation we obtain

$$H_{coll} = \frac{1}{2}\int dx\,[\phi^0(x)\,\xi^2(x) + \frac{1}{2}\frac{(\partial\eta)^2}{\phi^0} - \omega\frac{\eta^2}{\phi^0}] \tag{7.40}$$

The fluctuation $\eta(x)$ is subject to the following constraint:

$$\int dx\,\eta(x) = 0 \tag{7.41}$$

which is due to (7.33).

The normal mode decomposition of collective Hamiltonian (7.40) is straightforward. We quote the result.

$$\eta(x) = \phi^0(x)^{1/2} \sum_{n \neq 0} \sqrt{2\omega n}\;\chi_n(x) q_n \tag{7.42}$$

$$\xi(x) = \phi^0(x)^{-1/2} \sum_{n \neq 0} \frac{1}{\sqrt{2\omega n}}\chi_n(x) p_n \tag{7.43}$$

$$H_{coll} = \frac{1}{2}\sum_{n \neq 0}(p_n^2 + \omega_n^2 q_n^2) \tag{7.44}$$

where $\chi_n(x)$'s are the normalized eigenfunctions of harmonic oscillators:

$$\left[-\frac{1}{2}\partial_x^2 + \frac{1}{2}\omega^2 x^2\right]\chi_n(x) = (n + 1/2)\,\omega\,\chi_n(x) \tag{7.45}$$

The absence of $n=0$ mode in (7.44) is due to the constraint (7.41) (Note $\phi^0 \propto \chi_0$).

If we try to interpret the collective modes in terms of the classical motion of harmonic oscillators, they should correspond to the following classical collective motions: (i) All the particles are at $x=0$ for $t=0$ with positive initial velocities (see Fig. 7-1a). This corresponds to the $n=1$ collective motion. (ii) All the particles are at $x=0$ at $t=0$ with zero averaged initial velocity (see Fig. 7-1b). This corresponds to the $n=2$ collective motion.

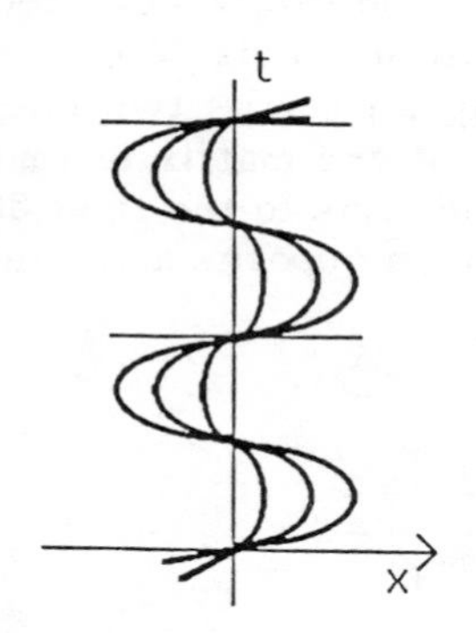

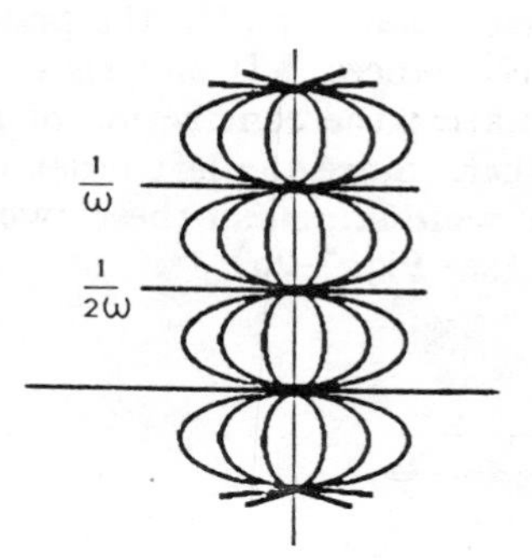

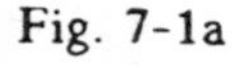
Fig. 7-1a

Fig 7-1b

7.2 Planar Limit of SU(N) Symmetric Hermitian Matrix Model

Planar Limit

The SU(N) symmetric Hermitian matrix system is defined by the following Lagrangian:

$$L = \frac{1}{2} tr\dot{M}^2 - \frac{1}{2} trM^2 - \frac{g}{N} trM^4 \tag{7.46}$$

where M is an N x N Hermitian matrix and the components of the matrix are the dynamical variables of the system. The quantization of this system is discussed in the Appendix (see **A2.2**).

Lagrangian (7.46) is invariant by SU(N) transformation:

$$M \to uMu^{+} \qquad u \in SU(N) \tag{7.47}$$

We consider only the singlet states. The Schrödinger wave function satisfies

$$\psi[M] = \psi[uMu^{+}] \tag{7.48}$$

As shown in **II** one can express the transition amplitudes by Feynman path integral and one can develop a perturbation expansion according to the procedure described in **IV** , where each term can be described by Feynman diagrams. In the present case, however, a matrix M carries two SU(N) indices, left and right, so that a propagator carries two lines each indicating the contraction of an SU(N) index of one matrix to that of the other. Since the left index of one matrix contracts to the right of the other we distinguish these two lines by adding an opposite arrow to each line (see Fig. 7-2a).

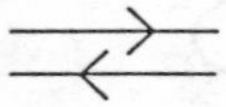

Fig. 7-2a

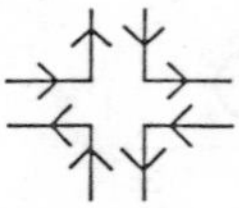

Fig. 7-2b

Similarly the vertex is expressed as in Fig. 7-2b. We call these lines the color lines anticipating the application of similar technique to QCD.

The singlet to singlet SU(N) invariant transition amplitudes are expressed in terms of Feynman diagrams in which all the color lines are connected. There should not be any free end of color lines. Each Feynman diagram has a number of color line loops. Each loop corresponds mathematically to $tr\ 1$ so that it is proportional to N. Thus, the N dependence of each Feynman diagram is given by N^{l} where l is the number of color loops in the diagram.

For example, the vacuum to vacuum diagrams of order g are given by Fig. 7-3a,b. One is planar and the other non-planar. The color lines are denoted by Fig. 7-4a,b. As one sees in these diagrams, the planar diagram contains 3 color loops while the non-planar diagrams contain one loop, so in the large N limit the planar diagram dominates. This is so in general. It is easy to convince ourselves now that the large N limit of any SU(N) symmetric matrix model (of course this includes the SU(N) gauge theory) is the planar limit.

Fig. 7-3a

Fig. 7-3b

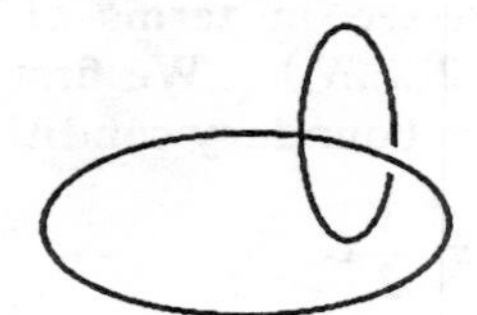

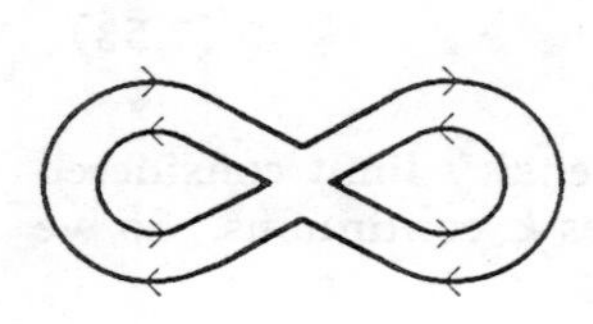

Fig. 7-4a

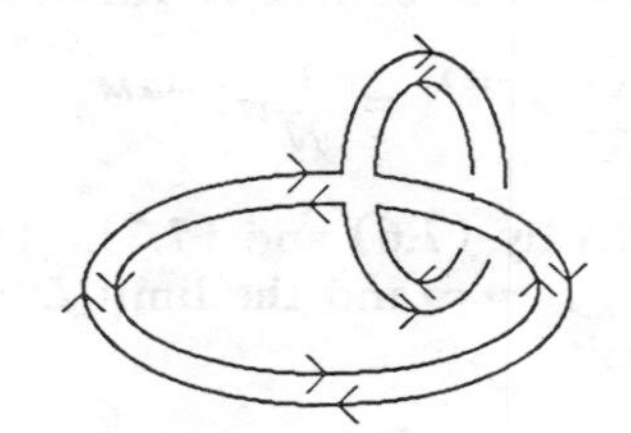

Fig. 7-4b

Collective Field Theory

The Hamiltonian of the system is given by

$$H = -\frac{1}{2}\sum_{a=0}^{N^2-1}\frac{\partial^2}{\partial M^{a2}} + tr\left(\frac{1}{2}M^2 + \frac{g}{N}M^4\right) \tag{7.49}$$

where

$$M = \sum_{a=0}^{N^2-1} t^a M_a \tag{7.50}$$

$t^a\,(a=0,1,2,\ldots,N^2-1)$ is the fundamental representation of U(N) Lie algebra which has the following property:

$$\sum_a (t^a)_{\alpha\beta}(t^a)_{\alpha'\beta'} = \delta_{\alpha\beta'}\delta_{\alpha'\beta} \tag{7.51}$$

They are an N dimensional extension of Pauli spin matrices.

Since we restrict the states to be singlets, we choose collective variables which are invariant by SU(N) transformation (7.47).

There are N independent invariants in this system. They are most conveniently expressed in terms of real eigenvalues of M, which we denote by $\lambda_i (i=1,2,\ldots,N)$. We first restrict the range of λ_i to be finite and impose periodic boundary conditions.

$$\frac{1}{2}L \geqslant \lambda_i \geqslant -\frac{1}{2}L \tag{7.52}$$

Later we set $L \to \infty$. Therefore, for the large N limit we must consider two limits: $N \to \infty$ and $L \to \infty$.

We choose a set of collective variables as before

$$\Phi_k = \frac{1}{N}\sum_{i=1}^{N} e^{-ik\lambda_i} = \frac{1}{N} tr e^{-ikM}. \tag{7.53}$$

where k is given by (7.6) and (7.7). The high density limit considered before makes $k_{max} \to \infty$ and the limit $L \to \infty$ makes k continuous. So we consider the limit

$$N/L \to \infty, \qquad L \to \infty \tag{7.54}$$

We then calculate ω and Ω

Calculation of ω :

$$\omega(k;\Phi) \equiv -\sum_a \frac{\partial^2 \Phi_k}{\partial M^{a2}} = -\frac{1}{N}\sum_a \frac{\partial^2}{\partial M^{a2}} tr\ e^{-ikM}$$

$$= \frac{k^2}{N}\int_0^1 d\alpha \sum_a tr(t^a e^{-i\alpha kM} t^a e^{-i(1-\alpha)kM})$$

$$= \frac{k^2}{N}\int_0^1 d\alpha tr(e^{-i\alpha kM})\, tr(e^{-i(1-\alpha)kM})$$

$$= Nk^2 \int_0^1 d\alpha \Phi_{\alpha k}\, \Phi_{(1-\alpha)k} \tag{7.55}$$

where

$$\frac{\partial}{\partial M^a} e^{-ikM} = -ik\int_0^1 d\alpha e^{-i\alpha kM} t^a e^{-i(1-\alpha)kM}$$

and (7.57) are used in this calculation (see (8.23)).

Calculation of Ω

$$\Omega(k_1, k_2, \Phi) \equiv \sum_a \frac{\partial \Phi_{k_1}}{\partial M^a} \frac{\partial \Phi_{k_2}}{\partial M^a}$$

$$= -\frac{k_1 k_2}{N^2} tr(t^a e^{-ik_1 M}) tr(t^a e^{-ik_2 M})$$

$$= -\frac{k_1 k_2}{N^2} tr(e^{-i(k_1+k_2)M})$$

$$= -\frac{k_1 k_2}{N} \Phi_{k_1+k_2} \tag{7.56}$$

We note that ω is of order N while Ω is of order N^{-1}.

Next, in order to make the effective Hamiltonian to a manageable form we introduce Fourier transformations:

$$\Phi(X) = \int \frac{dk}{2\pi} e^{ikX} \Phi_k = \frac{1}{N} tr\, \delta(X - M) \tag{7.57}$$

The Fourier transformation of ω and Ω are given by

$$\omega(X ; \Phi) = 2N \frac{\partial}{\partial X} [\Phi(X) G(X ; \Phi)] \tag{7.58}$$

$$\Omega(X, X'; \Phi) = \frac{1}{N} \frac{\partial}{\partial X} \frac{\partial}{\partial X'} [\delta(X - X') \Phi(X)] \tag{7.59}$$

where

$$G(X ; \Phi) = P \int \frac{\Phi(X')}{X - X'} dX' \tag{7.60}$$

and P is the symbol for the principal part of the singular integration. We remark here the range of X is given by

$$\frac{1}{2} L \geqslant X \geqslant -\frac{1}{2} L \tag{7.61}$$

with the understanding that the limit $L \to \infty$ is taken eventually.

Calculation of V_{coll}

The equation which corresponds to (6.20) is now given by

$$2\int dX'\Omega(X,X';\Phi)\,C(X') + \omega(X:\Phi) = 0 \tag{7.62}$$

Substituting (7.58) and (7.59) we obtain

$$\frac{2}{N}\frac{\partial}{\partial X}\int dX'[\frac{\partial}{\partial X'}\delta(X-X')\,\Phi(X)]\,C(X') + 2N\frac{\partial}{\partial X}[\Phi(X)\,G(X:\Phi)] = 0 \tag{7.63}$$

The integral in the first term is given by $-\Phi(X)\,\partial_X C(X)$. So we obtain

$$\partial_X C(X) = N^2 G(X:\Phi) \tag{7.64}$$

Using (7.59) we obtain

$$\begin{aligned} V_{coll} &\equiv \frac{1}{2}\sum C\,\Omega C \\ &= \frac{1}{2N}\int dX\,(\partial_X\, C(X))^2\,\Phi(X) \\ &= \frac{N^3}{2}\int dX\,\Phi(X)\,G^2(X;\Phi) \\ &= \frac{N^3\pi^2}{6}\int dX\,\Phi^3(X) \end{aligned} \tag{7.65}$$

The passage to the last expression in (7.65) would require proof. Let

$$f(z) = \frac{1}{\pi}\int dX\frac{\Phi(X)}{X-z} \tag{7.66}$$

$f(z)$ is an analytic function on the complex plane with a cut on the real axis from $-L/2$ to $+L/2$. Since $\Phi(X)$ is constrained by

$$\int dX\,\Phi(X) = 1 \tag{7.67}$$

the function $f(z)$ approaches at

$$f(z) \rightarrow -\frac{1}{\pi}\frac{1}{z} \tag{7.68}$$

at $z=\infty$. Therefore, for the contour C at infinity we have

$$\int_C f^3(z)dz = 0 \tag{7.69}$$

We then deform the contour to the real axis and we obtain

$$\int dX(f^3(X+i\epsilon) - f^3(X-i\epsilon)) = 0 \tag{7.70}$$

Since

$$f(X+i\epsilon) = -\frac{1}{\pi}G(X;\Phi) + i\Phi(X) \tag{7.71}$$

Note

$$(X' - X - i\epsilon)^{-1} = P(X' - X)^{-1} + i\pi\delta(X' - X)$$

we obtain

$$\frac{3}{\pi^2}\int dXG^2(X;\Phi)\Phi(X) - \int dX\,\Phi^3(X) = 0 \tag{7.72}$$

(QED)

The calculation of K and $\tilde{V}$ is straightforward and does not require any further explanation. We obtain the following effective Hamiltonian:

$$H_{eff} = \int_{-L/2}^{L/2} dX[\frac{1}{2N}(\partial_x \Pi(X))^2\Phi(X) + \frac{N^3\pi^2}{6}\Phi^3(X)$$

$$+ N(\frac{1}{2}X^2 + \frac{g}{N}X^4)\Phi(X)] \tag{7.73}$$

where $\Pi(X)$ is the canonical conjugate to $\Phi(X)$:

$$[\Pi(X), \Phi(X')] = -i(\delta(X - X') - const) \tag{7.74}$$

In (7.73) we explicitly remain the dependence of L and N. However, the limit (7.54) is to be understood.

Large N Limit

We make a scale transformation such that all the terms in H_{eff} are the same order in N.

$$X = \sqrt{N}\,x \tag{7.75}$$

$$\Phi(X) = \frac{1}{\sqrt{N}}\,\phi(x), \qquad \Pi(X) = N^2\pi(x) \tag{7.76}$$

We obtain

$$H_{eff} = N^2\tilde{H}_{eff} = N^2(\tilde{K} + \tilde{V}_{eff}) \tag{7.77}$$

$$\tilde{K} = ½\int dx\,(\partial_x \pi(x))^2\,\phi(x) \tag{7.78}$$

$$\tilde{V}_{eff} = \int dx \left| \frac{\pi^2}{6}\phi^3(x) + (\frac{1}{2}x^2 + gx^4)\,\phi(x) \right| \tag{7.79}$$

The range of x is now given by

$$\frac{L}{2\sqrt{N}} \geqslant x \geqslant -\frac{L}{2\sqrt{N}} \tag{7.80}$$

and the constraint condition and the commutation relation are given by

$$\int dx\ \phi(x) = 1 \tag{7.81}$$

$$[\pi(x)\,.\,\phi(x')] = -\frac{i}{N^2}(\delta(x-x') - const.) \tag{7.82}$$

Since the commutator is of order N^{-2}, in the large N limit the dynamics of H_{eff} becomes classical. We take $L \to \infty$ so that

$$\infty \geqslant x \geqslant -\infty \tag{7.83}$$

The solution of the classical equation of motion derived from the effective Hamiltonian (7.73) is the expectation value of $\hat{\phi}(x)$ in the ground state.

$$\phi^0(x) = <\hat{\phi}(x)>_0 \tag{7.84}$$

It gives the distribution of the eigenvalues in the ground state. The equation for $\phi^0(x)$ is obtained by minimizing

$$V_{eff} = N^2\int dx\ \phi(x)\left| \frac{\pi^2}{6}\phi^2(x) + ½x^2 + gx^4 \right| - N^2\epsilon(\int dx\,\phi(x) - 1) \tag{7.85}$$

The last term is the Lagrange multiplier term. The solution is given by

$$\phi^0(x) = \frac{1}{\pi}(2\epsilon - x^2 - 2gx^4)^{1/2} \qquad \text{for } |x| < \Lambda$$

$$= 0 \qquad \text{for } |x| > \Lambda \qquad (7.86)$$

where Λ satisfies

$$2\epsilon - \Lambda^2 - 2g\Lambda^4 = 0 \qquad (7.87)$$

The Lagrange multiplier ϵ is determined by

$$\int dx\, \phi^0(x) = 1 = \frac{1}{\pi}\int_{-\Lambda}^{\Lambda}(2\epsilon - x^2 - 2gx^4)^{1/2} dx \qquad (7.88)$$

The ground state energy is obtained by inserting this solution into V_{eff}.

$$E_0(g) = V_{eff}[\phi^0] = N^2\left\{\epsilon - \frac{1}{3\pi}\int_{-\Lambda}^{\Lambda} dx\,(2\epsilon - x^2 - 2gx^4)^{3/4}\right\} \qquad (7.89)$$

Exercise: Show that for the strong coupling E_0 is given by

$$E_0(g)/N^2 \sim \frac{3}{7}\left[\frac{3}{2\sqrt{\pi}}\Gamma^2(3/4)\right]^{4/3} g^{1/3} = 0.58993 g^{1/3} \qquad (7.90)$$

Collective Excitations:

By setting

$$\phi(x) = \phi^0(x) + \eta(x) \qquad (7.91)$$

we expand H_{eff} in powers of η:

$$H_{eff} = E_0(g) + H_{coll} + \cdots \qquad (7.92)$$

H_{coll} is quadratic in η and π. We obtain

$$H_{coll} = \frac{N^2}{2}\int dx\, \phi^0(x)[(\partial_x \pi(x))^2 + \frac{\pi^2}{2}\eta^2(x)]$$

$$= \frac{1}{2}\int dx\, \phi^0(x)[(\partial_x \tilde{\pi}(x))^2 + \frac{\pi^2}{2}\tilde{\eta}^2(x)] \qquad (7.93)$$

where

$$\tilde{\pi} = N\pi, \qquad \tilde{\eta} = N\eta \tag{7.94}$$

so that

$$[\tilde{\pi}(x), \tilde{\eta}(x')] = -i(\delta(x - x') - const) \tag{7.95}$$

$$\int dx\; \tilde{\eta}(x) = 0 \tag{7.96}$$

Since $\eta = \tilde{\eta}/N$ and $\tilde{\pi}$ and $\tilde{\eta}$ are of order 1, the expansion (7.91) and (7.92) are $1/N$ expansions.

The normal mode expansion of H_{coll} is more involved. We simply quote the results.

$$\xi(x) = \int_{-\Lambda}^{x} \frac{dy}{\phi^0(y)}, \qquad T = \int_{-\Lambda}^{\Lambda} \frac{dy}{\phi^0(y)} \tag{7.97}$$

$$\tilde{\eta}(x) = \frac{1}{\phi^0(x)} \sum_{n=1}^{\infty} (2/T)^{1/2} \cos(n\pi\xi(x)/T)(n\pi/T)\, q_n \tag{7.78}$$

$$\tilde{\pi}(x) = -\frac{1}{\phi^0(x)} \sum_{n=1}^{\infty} (2/T)^{1/2} \sin(n\pi\xi(x)/T)\, p_n \tag{7.99}$$

$$[p_n, q_{n'}] = -i\,\delta_{nn'} \tag{7.100}$$

$$H_{coll} = \frac{1}{2} \sum_{n=1}^{\infty} (p_n^2 + \omega_n^2(g)\, q_n^2) \tag{7.101}$$

$$\omega_n^2(g) = n^2\pi^2/T \tag{7.102}$$

The excitations of order 1 are therefore given by these harmonic excitations.

VIII. Variational Method

The non-perturbative approximation methods in quantum mechanics discussed in the standard text books of quantum mechanics are the variational method and the WKB (semi-classical) method. In the next several lectures we discuss these subjects for many variables systems and field theories.

In the standard treatment of the variational method one uses trial wave function in Scrödinger formalism. Although we used a concept of wave function rather extensively in this lecture, in the standard treatment of field theory the wave function appears only implicitly. Therefore, in this lecture intending the application to field theories we discuss the variational method in path integral formalism. The method is due to Feynman.

8.1 Feynman's Variational Method

Let us consider a problem of obtaining the ground state energy E_0 of a Hamiltonian system. We assume H has a standard form:

$$\hat{H} = \frac{1}{2}\hat{p}^2 + V(\hat{q})$$

We first consider the partition function defined by

$$Z(\beta) \equiv e^{-W(\beta)} = tr(e^{-\beta\hat{H}}) = \sum_n e^{-\beta E_n} \tag{8.1}$$

then the ground state energy is given by

$$E_0 = \lim_{\beta\to\infty} W(\beta)/\beta \tag{8.2}$$

The partition function has the following path integral representation:

$$Z(\beta) = \int\cdots\int DpDq\; e^{-\int_0^\beta(-ip\dot{q} + H(p,q))d\tau} \tag{8.3}$$

$$= \int\cdots\int Dq\; e^{-\int_0^\beta(\frac{1}{2}\dot{q}^2 + V(q))d\tau}$$

$$\equiv \int \cdots \int Dq \; e^{-S[q]} \tag{8.4}$$

The inequality

$$\frac{1}{2}(e^{-x_1} + e^{-x_2}) \geqslant e^{-\frac{1}{2}(x_1 + x_2)} \tag{8.5}$$

can be easily read from Fig. 8-1. This is a special case of the Jensen inequality:

$$\int d\vec{x} f\,(g(\vec{x}))P(\vec{x}) \geqslant f\,(\int d\vec{x} g\,(\vec{x})P(\vec{x})) \tag{8.6}$$

where f is a convex function of a real variable, i.e. $f'' > 0$, $g(\vec{x})$ a real function of many variables $(x_1, x_2, \cdots x_n)$ and $P(\vec{x})$ a positive function.

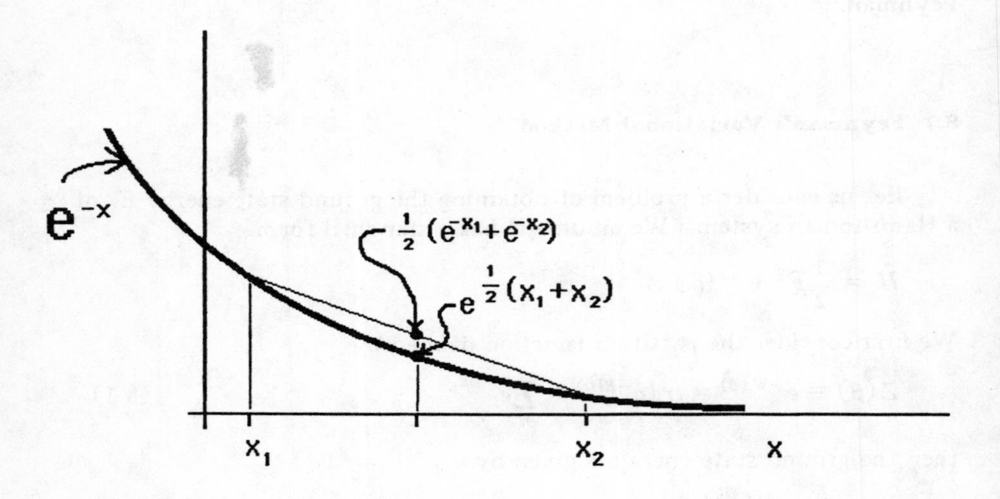

Fig. 8-1

Define the average of $F(q)$ by

$$< F >_0 = \frac{\int \cdots \int D\, qF(q)\, e^{-S_0[q]}}{\int \cdots \int Dq \; e^{-S_0[q]}} \tag{8.7}$$

then,

$$Z(\beta) = \int \cdots \int Dq\ e^{-(S[q] - S_0[q])}\ e^{-S_0[q]}$$

$$= \langle e^{-(S[q] - S_0[q])} \rangle_0\ e^{-W_0(\beta)} \tag{8.8}$$

where,

$$e^{-W_0(\beta)} = \int \cdots \int Dq e^{-S_0[q]} \tag{8.9}$$

Applying the Jensen inequality one obtains

$$\langle e^{-S[q] - S_0[q]} \rangle_0 \ \geqslant e^{-\langle S - S_0 \rangle_0} \tag{8.10}$$

Therefore, if we define

$$W_{eff} \ \equiv \langle S - S_0 \rangle_0 + W_0(\beta) \tag{8.11}$$

then,

$$E_{eff} \ \equiv \lim_{\beta\to\infty} W_{eff} \,/\, \beta \ \ \geqslant \ \lim_{\beta\to\infty} W(\beta)/\ \beta = E_0 \tag{8.12}$$

gives an upper bound of E_0 .

We call S_0 a trial action, which may contain parameters $\kappa_1 .... \kappa_s$. Then E_{eff} is a function of κ's so that one can minimize E_{eff} by the variation:

$$\frac{\partial E_{eff}}{\partial \kappa_i} = 0 \tag{8.13}$$

We remark that the parameters appeared in the trial action can be time dependent, since we used the Jensen inequality directly to the path integral (8.4). However, the path integral (8.4) assumes the standard form of Hamiltonian. When Hamiltonian does not have the standard form, we should develop a variational method based on the phase space path integral (8.3). Unfortunately, however, the Euclidean action which appeared in (8.3) is complex so that one can not apply the Jensen inequality to this integral.

To develop the variational method in phase space path integral let us introduce a trial Hamiltonian, which depends on a set of *time-independent* parameters, say $\kappa_1 ..... \kappa_s$, and consider the partition function

$$Z_0(\beta) \equiv e^{-W_0(\beta)} = tr e^{-\beta\hat{H}_0} \tag{8.14}$$

We define W_{eff} analogous to (8.11) by

$$W_{eff} \equiv \beta < \hat{H} - \hat{H}_0 >_0 + W_0(\beta) \tag{8.15}$$

where

$$< F >_0 = \frac{trFe^{-\beta\hat{H}_0}}{tre^{-\beta H_0}} \tag{8.16}$$

The inequality one can use is the Gibbs inequality

$$W_{eff}(\beta) \geqslant W(\beta). \tag{8.17}$$

which we shall prove later. Since average (8.16) is expressed by an analogous average in phase space path integral (see (2.13) for the relation between operator expression and path integral) the same variational method described above is applicable for the phase space path integral provided that the trial action contains a set of *time-independent* parameters only.

Proof of (8.17):

We follow Feynman's proof. Let

$$Z(\alpha,\beta) \equiv e^{-W(\alpha,\beta)} = tre^{-\beta\hat{H}(\alpha)} \tag{8.18}$$

where

$$\hat{H}(\alpha) = \hat{H}_0 + \alpha(\hat{H} - \hat{H}_0) \tag{8.19}$$

Then

$$W(\beta) = W(1,\beta), \qquad W_0(\beta) = W(0,\beta). \tag{8.20}$$

Next, we prove

$$\left.\frac{\partial W}{\partial \alpha}\right|_{\alpha=0} = \beta < \hat{H} - \hat{H}_0 >_0 \tag{8.21}$$

and

$$\frac{\partial^2 W}{\partial \alpha^2} \leqslant 0 \tag{8.22}$$

for all α. Then, $W(\alpha,\beta) \leqslant W(0,\beta) + \alpha \left(\frac{\partial W}{\partial \alpha}\right)_{\alpha=0}$ so by setting $\alpha = 1$ we obtain (8.17).

We use

$$\frac{\partial}{\partial\alpha}e^{\hat{A}} = \int_0^1 d\gamma\, e^{\gamma\hat{A}} \frac{\partial\hat{A}}{\partial\alpha} e^{(1-\gamma)\hat{A}} \tag{8.23}$$

then (8.21) follows. The proof of (8.22) is more complicated. Using (8.23)

$$\frac{1}{\beta^2}\frac{\partial^2 W(\alpha,\beta)}{\partial\alpha^2} = (<\hat{H} - \hat{H}_0>_\alpha)^2 -$$

$$- <(\hat{H} - \hat{H}_0)\int_0^1 d\gamma\, e^{-\beta\gamma\hat{H}(\alpha)}(\hat{H} - \hat{H}_0)e^{+\beta\gamma\hat{H}(\alpha)}>_\alpha$$

$$= \left(\frac{\sum_n <n|(\hat{H} - \hat{H}_0)|n>e^{-\beta E_n}}{\sum_n e^{-\beta E_n}}\right)^2 - \sum_n e^{-\beta E_n}(<n|(\hat{H} - \hat{H}_0)|n>)^2 -$$

$$- \sum_{m\neq n} e^{-\beta E_n}\frac{e^{\beta(E_n - E_m)} - 1}{E_n - E_m}|<n|(\hat{H} - \hat{H}_0)|m>|^2 \tag{8.24}$$

where E_n 's are eigenvalues of $\hat{H}(\alpha)$. The first two terms are negative since

$$\sum_n f(g_n)P_n \geqslant f(\sum_n g_n P_n) \tag{8.25}$$

which is a discrete version of Jensen inequality (8.6). The last term is obviously negative since the function $\frac{1}{\kappa}(e^\kappa - 1)$ is positive for all real κ

Exercise: Prove (8.23)

8.2 Lee Low Pines Theory of Polaron

Polaron Problem:

We study the motion of an electron in a polar crystal. The Hamiltonian for this system has been given by Fröhlich and has the form

$$H = \frac{\hat{p}^2}{2m} + \sum_{\vec{k}} \omega(\hat{a}_k^+ \hat{a}_k + \frac{1}{2}) + \sum_{\vec{k}}(V_k \hat{a}_k e^{i\vec{k}\cdot\vec{r}} + h.c.) \qquad (8.26)$$

where $\hat{r}$ and $\hat{p}$ are the position and momentum of the electron, $\hat{a}_k^+$ and $\hat{a}_k$ are the creation and annihilation operators of a phonon of mode k and V_k the phonon-electron coupling given by

$$V_k = \frac{i}{k}\,\omega \left[\frac{1}{2\omega m}\right]^{1/4} \left[\frac{4\pi\alpha}{\Omega}\right]^{1/2} \qquad (8.27)$$

In this expression, Ω is the volume, α is the real coupling between electron and phonons given by

$$\alpha = \frac{1}{2}\,(\frac{1}{\epsilon_\infty} - \frac{1}{\epsilon_0})\frac{e^2}{\hbar\omega}\,(\frac{2m\omega}{\hbar})^{1/2} \qquad (8.28)$$

and ω is the frequency of the phonon which we assume independent of k, ϵ_∞ and ϵ_0 are the values of the dielectric constant at $\omega = \infty$ and $\omega = 0$ respectively.

In the case $m = \infty$ the kinetic energy of the electron is zero so that $\hat{r}$ is a constant of motion (static). In this case the Hamiltonian (8.26) becomes equivalent to a static model of neutral scalar meson, which is exactly soluble. However, in the polaron problem the motion of the electron is essential and one can not use static approximation.

Another manner to write $\hat{H}$ is to introduce a real scalar field

$$\hat{\phi}(\vec{x}) = \frac{i}{\sqrt{2\omega\Omega}} \sum_{\vec{k}} (\hat{a}_k e^{i\vec{k}\cdot\vec{x}} - \hat{a}_k^+ e^{-i\vec{k}\cdot\vec{x}}) \qquad (8.29)$$

$\hat{\phi}(x) = \hat{\phi}^+(x)$. The Hamiltonian (8.26) is then written as

$$\hat{H} = \frac{\hat{p}^2}{2m} + \frac{1}{2}\int d\vec{x}\ (\hat{\pi}^2(\vec{x}) + \omega^2\hat{\phi}^2(\vec{x})) + \int d\vec{x}\,\rho(\vec{x} - \hat{r}))\,\phi(\vec{x}) \qquad (8.30)$$

where $\rho(\vec{x})$ is the Fourier transform of V_k .

$$V_k = \frac{i}{\sqrt{2\omega\Omega}} \int e^{i\vec{k}\cdot\vec{x}}\,\rho(\vec{x})d\vec{x} = \frac{i}{\sqrt{2\omega}}\rho(-k) \qquad (8.31)$$

Change of Variables:

We see that in (8.30) $\hat{\vec{r}}$ enters only through $\rho(\vec{x} - \hat{\vec{r}})$. The $\hat{r}$

dependence can be eliminated if one performs the following canonical transformation:

$$\vec{r}(t) \quad \to \quad \vec{r}(t)\,,$$

$$\phi(\vec{x},t) \quad \to \quad \xi(\vec{x},t) = \phi(\vec{x} + \vec{r}(t), t) \tag{8.32}$$

This is a transformation from the laboratory frame to the moving frame of the electron. Namely, the field $\xi(\vec{x},t)$ is the phonon field in the moving frame. As we learned in **VI** , when we make a canonical transformation we have to add an extra potential ΔV that is given by

$$\Delta V = \frac{1}{8m}\Gamma^i_{jk} \;\; \Gamma^j_{il}\, g^{kl} \tag{8.33}$$

Our transformation is (8.32) and the variables are $\vec{r}$ and $\xi(\vec{x})$. We specify them by index 0 for $\vec{r}$ and by index x for $\xi(\vec{x})$. We then obtain

$$g_{00} = 1 + \int dx\,(\frac{\partial}{\partial r}\xi(x - r))\cdot(\frac{\partial}{\partial r}\xi(x - r))$$

$$= 1 + \int dx\,(\xi'(x))^2$$

$$g_{0x} = \int dy \frac{\partial\xi(y - r)}{\partial r}\;\frac{\partial\xi(y - r)}{\partial\xi(x)} = -\,\xi'(x)$$

$$g_{xy} = \delta(x - y)$$

where

$$\xi'(x) \equiv \nabla\xi(x)$$

The inverse of g_{ab} .g^{ab} is given by

$$g^{00} = 1$$

$$g^{0x} = \xi'(x)$$

$$g^{xy} = \delta(x - y) + \xi'(x)\xi'(y) \tag{8.34}$$

Putting all this in (8.33) we get

$$\Delta V = -\frac{1}{8m}\int dxdy\,(\delta'(x - y))^2 \tag{8.35}$$

The new Hamiltonian is

$$H = \frac{1}{2m}(\vec{p} + \int d\vec{x}\, \pi \vec{\nabla} \xi)^2 + \int d\vec{x}(\frac{1}{2}\pi^2 + \frac{1}{2}\omega^2\xi^2 + \rho\xi) + \Delta V \quad (8.36)$$

The two terms in the first parenthesis and the first term in the second parenthesis come from $\frac{1}{2}(p_a g^{ab} p_b)_W$.

We see immediately that r is a cyclic variable so p is conserved, which is the total momentum of the system, i.e. the momentum of the dressed electron.

We remember that

$$Z(\beta) = \int \cdots \int DpDrD\pi D\xi e^{-S} \quad (8.37)$$

where action S is given by

$$S = \int_0^\beta d\tau(-i\vec{p}\cdot\dot{\vec{r}} - i\pi\dot{\xi} + H)$$

as $\vec{r}$ appears only in the first term, we can integrate it out in (8.37) and we are left with

$$Z(\beta) = \int \cdots \int D\pi D\xi Dp\, \delta(\dot{p})e^{-S}$$

where

$$S = \int_0^\beta (-i\pi\dot{\phi} + H)d\tau$$

We introduce the partition function for fixed momentum:

$$Z(\beta,p) = \int \cdots \int D\pi D\phi\, e^{-S} \equiv e^{-W(\beta,p)} \quad (8.38)$$

then

$$Z(\beta) = \int dpZ(\beta,p) \quad (8.39)$$

The ground state energy of the system (polar crystal and an electron) for a fixed momentum is then obtained by

$$E_0(p) = \lim_{\beta\to\infty} W(\beta,p)/\beta \quad (8.40)$$

Making an expansion of $E_0(p)$ around $p=0$, we expect

$$E_0(p) = E_0 + \frac{\vec{p}^2}{2m^*} + \ldots \quad (8.41)$$

where m^* is an "effective mass" for the electron in the crystal and E_0 is the ground state energy of the crystal. We intend to calculate E_0 and m^*.

Variational Method Applied to the Polaron Problem:

We are now ready to apply the variational method. We choose the trial action

$$S_0 = A_0[\pi - \pi_c, \xi - \xi_c] \tag{8.42}$$

where A_0 is the free action given by

$$A_0 = \int_0^\beta d\tau \int d\vec{x} \, [-i\pi\dot{\xi} + \frac{1}{2}(\pi^2 + \omega^2\xi^2)] \tag{8.43}$$

We treat π_c and ξ_c as variational parameters, which we assume time independent as discussed in **8.1** .

Looking at formula (8.11) we see that the first to be calculated is

$<S - S_0> =$

$$= \frac{\int \cdots \int D\pi D\xi \, (S[\pi,\xi] - S_0[\pi,\xi]) \, e^{-S_0[\pi,\xi]}}{\int \cdots \int D\pi D\xi e^{-S_0[\pi,\xi]}} \tag{8.44}$$

Making a shift of variables in the integral, we obtain

$<S - S_0 > =$

$$= \frac{\int \cdots \int D\pi D\xi \, (S[\pi + \pi_c, \xi + \xi_c] - A_0[\pi,\xi]) \; e^{-A_0[\pi,\xi]}}{\int \cdots \int D\pi D\xi \; e^{-A_0[\pi,\xi]}} \tag{8.45}$$

The evaluation of this integral can be done by the standard perturbation technique as the average is made with the free action A_0 .

Let us now analyze

$<S[\pi + \pi_c, \xi + \xi_c] - A_0[\pi,\xi]>_0$

Since $A_0[\pi,\xi]$ is invariant under $\pi \to -\pi$, $\xi \to -\xi$ we can drop all the odd terms in π and ξ . We are left with the average of

$S[\pi_c, \xi_c] +$

$$+ < \frac{1}{2m} \Big| \pi^2(\vec{\nabla}\xi)^2 + \pi_c^2(\vec{\nabla}\xi)^2 + \pi^2(\vec{\nabla}\xi_c)^2 +$$

$$+ 4\pi\, \pi_c (\vec{\nabla}\xi)(\vec{\nabla}\xi_c) + 2\vec{p}\,\pi\cdot\vec{\nabla}\xi \Big| >_0 \quad (8.46)$$

where $\int_0^\beta d\tau \int d\vec{x}$ is understood in all terms. We can now use the rules derived in **VI** that

$$<\xi(\vec{x})\xi(\vec{x}')>_0 = \delta(\vec{x} - \vec{x}')\Delta(\tau - \tau')$$

$$<\pi(\vec{x})\pi(\vec{x}')>_0 = \omega^2\delta(\vec{x} - \vec{x}')\Delta(\tau - \tau')$$

$$<\pi(\vec{x})\xi(\vec{x}')>_0 = i\,\dot{\Delta}(\tau - \tau')\delta(\vec{x} - \vec{x}')$$

$$\Delta(\tau) = \frac{1}{\beta}\sum_n \frac{1}{\omega^2 + (\frac{2\pi n}{\beta})^2} e^{\frac{2\pi i n \tau}{\beta}} \quad (8.47)$$

therefore,

$$\dot{\Delta}(0) = 0$$

$$\Delta(0) = \frac{1}{2\omega} + 0(\frac{1}{\beta}) \quad (8.48)$$

and by this we see that the last two terms in (8.46) are zero, and

$$< \frac{1}{2m}(\int dx\, \pi\nabla\xi)^2 >_0 = \Delta V$$

$$< \frac{1}{2m}(\int \pi_c\, \nabla\xi)^2 + \int (\pi\nabla\xi_c)^2 >_0 = \frac{1}{4m\,\omega}\int [(\nabla\pi_c)^2 + \omega^2(\nabla\xi_c)^2]dx$$

So we get

$$<S - S_0>_0 = \int_0^\beta d\tau [\frac{1}{2m}(\vec{p} + \int \pi_c \vec{\nabla}\xi_c\, dx)^2 +$$

$$+ \int dx (\frac{1}{2}\pi_c^2 + \frac{1}{2}\omega^2\xi_c^2 + \rho\xi_c) + \frac{1}{4m\,\omega}\int dx ((\vec{\nabla}\pi_c)^2 + \omega^2(\vec{\nabla}\xi_c)^2)] \quad (8.49)$$

Looking back at formula (8.11) we see that we need W_0. This can be easily calculated in the following way.

$$Z_0 \equiv e^{-W_0} = \int \cdots \int D\pi D\xi\, e^{-S_0}$$

$$= \int \cdots \int D\pi D\xi \, e^{-A_0[\pi,\xi]}$$

$$= \int \cdots \int D\xi \, e^{-\int_0^\beta d\tau \int d\vec{x} \frac{1}{2}(\dot{\xi}^2 + \omega^2 \xi^2)}$$

$$= (\det K)^{-1/2} \tag{8.50}$$

where

$$<x',\tau' | K | x,\tau> = (-\frac{\partial^2}{\partial \tau^2} + \omega^2)\delta(\tau - \tau')\delta(x - x')$$

$$<x',\tau' | K^{-1} | x,\tau> = \delta(\vec{x} - \vec{x}')\Delta(\tau - \tau') \tag{8.51}$$

so,

$$(\det K)^{-1/2} = \sum_n e^{-\beta\omega(n + \frac{1}{2})\Omega} \tag{8.52}$$

and

$$W_0 \equiv \ln Z_0 = -\ln \sum_n e^{-\beta\omega(n + \frac{1}{2})\Omega} \tag{8.53}$$

where Ω is the spatial volume of the system.

Now we have all the elements for E_{eff}

$$E_{eff} = \lim_{\beta \to \infty} \frac{1}{\beta} \left| -\ln \sum_n e^{-\beta\omega(n + \frac{1}{2})\Omega} \right| + \lim_{\beta \to \infty} \frac{1}{\beta}(8.49)$$

The first term is the zero point energy of the phonons. In the second term, as π_c and ξ_c are independent of τ we can factor out β and we get

$$E_{eff} = \frac{1}{2m}(\vec{p} + \int dx\, \pi_c \vec{\nabla}\xi_c)^2 + \int dx\, (\frac{1}{2}\pi_c^2 + \frac{1}{2}\omega^2\xi_c^2 + \rho\xi_c) -$$

$$+ \frac{1}{4m\omega}\int dx\, [(\vec{\nabla}\pi_c)^2 + \omega^2(\vec{\nabla}\xi_c)^2] \tag{8.54}$$

Remember that π_c and ξ_c are variational parameters. We have to determine them by minimizing E_{eff} , as E_{eff} is an upper bound on E_0 . So, we get a system of differential equations for π_c and ϕ_c from

$$\frac{\delta E_{eff}}{\delta \xi_c} = 0, \qquad \frac{\delta E_{eff}}{\delta \pi_c} = 0 \tag{8.55}$$

Using (8.54) we obtain

$$\omega^2(1 - \frac{1}{2m\omega}\nabla^2)\xi_c - \frac{1}{m}(\vec{p} + \int \pi_c \vec{\nabla}\xi_c dx)\cdot\vec{\nabla}\pi_c + \rho = 0 \quad (8.56)$$

$$(1 - \frac{1}{2m\omega}\nabla^2)\pi_c + \frac{1}{m}(\vec{p} + \int \pi_c \vec{\nabla}\xi_c dx)\cdot\vec{\nabla}\xi_c = 0 \quad (8.57)$$

To solve this system we take Fourier transform

$$\xi_c(\vec{x}) = \frac{1}{\sqrt{\Omega}} \sum_{\vec{k}} e^{i\vec{k}\cdot\vec{x}} \xi(\vec{k}) \quad (8.58)$$

The reality of ξ_c and π_c implies

$$\xi(\vec{k}) = \xi^*(-\vec{k})$$

$$\pi(\vec{k}) = \pi^*(-\vec{k}) \quad (8.59)$$

Note that $\int \pi_c \vec{\nabla}\xi_c dx$ is a vectorial quantity. The only vector of the system is the momentum. Therefore, we take the ansatz:

$$-\int \pi_c \vec{\nabla}\xi_c d\vec{x} = \vec{p}\eta \quad (8.60)$$

where η is a constant to be determined later. (8.56) and (8.57) then become

$$\omega^2(1 + \frac{k^2}{2m\omega})\xi(\vec{k}) - \frac{i}{m}\vec{p}\cdot\vec{k}(1-\eta)\pi(\vec{k}) + \rho(\vec{k}) = 0$$

$$(1 + \frac{k^2}{2m\omega})\pi(\vec{k}) + \frac{i}{m}\vec{p}\cdot\vec{k}(1-\eta)\xi(\vec{k}) = 0 \quad (8.61)$$

We define

$$f_k = \frac{1}{\sqrt{2\omega}}(\omega\xi(\vec{k}) + i\pi(\vec{k})) \quad (8.62)$$

Multipling the second equation of (8.61) by $i\omega$ and adding to the first equations we obtain

$$f_k = \frac{iV^*_k}{\omega - \frac{1}{m}(\vec{p}\cdot\vec{k})(1-\eta) + \frac{k^2}{2m}} \quad (8.63)$$

where

$$V^*_k = -i\rho(\vec{k})/\sqrt{2\omega} \tag{8.64}$$

Now, we prove

$$\eta\vec{p} = \sum_{\vec{k}} \vec{k}\,|f_k|^2 \tag{8.65}$$

Indeed,

$$\eta\vec{p} = -\int d\vec{x}(\pi_c \vec{\nabla}\xi_c) = -i\sum_k \pi(-\vec{k})\vec{k}\,\xi(\vec{k}) \tag{8.66}$$

on the other hand,

$$\sum_k \vec{k}\,|f_k|^2 =$$

$$= \sum_{\vec{k}} \frac{\vec{k}}{2\omega}[\omega^2\xi(\vec{k})\xi(-\vec{k}) + \pi(\vec{k})\pi(-\vec{k}) + i\,2\omega\pi(\vec{k})\xi(-\vec{k})] \tag{8.67}$$

In this sum only the last term is not zero which gives precisely (8.65).

From (8.65) and (8.63) we obtain

$$\eta\vec{p} = \sum_k |V_k|^2 \vec{k} \frac{1}{(\omega - \frac{1}{m}\vec{p}\cdot\vec{k}(1-n) + \frac{k^2}{2m})^2} \tag{8.68}$$

Inserting these results into (8.54) we obtain

$$E = \frac{p^2}{2m}(1-\eta^2) + \sum_k \frac{|V_k|^2}{\frac{\vec{k}\cdot\vec{p}}{m}(1-\eta) - \frac{k^2}{2m} - \omega} \tag{8.69}$$

Insert here the explicit form V_k (8.27) and go over the integration over k using $\frac{1}{\Omega}\sum_k = \int \frac{d\vec{k}}{(2\pi)^3}$. We obtain

$$E = \frac{p^2}{2m}(1-\eta^2) - \alpha\omega\frac{\sin^{-1}Q}{Q} \tag{8.70}$$

where

$$Q = \frac{p(1-\eta)}{\sqrt{2\omega m}} \tag{8.71}$$

Next we calculate the effective mass m^* using the expression (8.41). We expand (8.70) in the power of p^2.

$$E = -\alpha\omega = \frac{p^2}{2m}[(1-\eta^2) - \frac{\alpha}{6}(1-\eta)^2] + 0(p^4) \tag{8.72}$$

Since η appears in the coefficient of p^2 , we can put p = 0 to determine η. From (8.68) , we obtain

$$\eta\vec{p} = (1-\eta)\sum_k \frac{|V_k|^2 \vec{k}\frac{2\vec{k}\cdot\vec{p}}{m}}{(\omega + \frac{k^2}{2m})^3} \tag{8.73}$$

and

$$\frac{\eta}{1-\eta} = \frac{8\alpha}{3\pi}\int_0^\infty \frac{x^2}{(1+x^2)^3}dx = \frac{\alpha}{6}$$

$$\eta = \frac{\alpha/6}{1+\frac{\alpha}{6}} \tag{8.74}$$

Insert this into the energy expression (8.69) and we get

$$E(p) = -\alpha\omega + \frac{p^2}{2m(1+\frac{\alpha}{6})} \tag{8.75}$$

Hence,

$$E_0 = -\alpha\omega, \qquad \frac{m^*}{m} = 1 + \frac{\alpha}{6} \tag{8.76}$$

Remarks:

The coupling strength α is 3.6 for *KCl* . It varies from 1 to 6 depending on the crystal. It is moderately strong so that the weak coupling perturbation is not good for the polaron problem.

There exist two different variational calculations for polaron problem Lee-Low-Pines' and Feynamn's. Lee-Low-Pines uses Tomonaga's intermediate coupling theory to this problem, which is the standard variational method extended to the fixed source meson theory. The key to the Tomonoga theory is a specific choice of the trial state function. On the other hand, Feynman uses Feynman's variational method described in this lecture. But, he uses it after all the ϕ integration is carried out.

Note the ϕ integration is a quadratic integral in this problem so that it can be done exactly. Since the phonon field is treated exactly in Feynman's calculation, it provides a more precise result.

The variational calculation presented here is an attempt to use Feynman's method to derive the result of Lee-Low-Pines. The original intention was to understand Tomonoga's intermediate coupling theory in path integral formalism.

8.3 Ground State Energy of the SU(N) Symmetric Hermitian Matrix Model

In the previous chapter we discussed the large N limit of the SU(N) symmetric matrix model. Here we evaluate the ground state energy for the same system in terms of the variational method developed in the previous sections.

The Hamiltonian of the system is given by

$$H = -\frac{1}{2}\frac{\partial^2}{\partial M^{a2}} + \frac{1}{2}trM^2 + \frac{g}{N}trM^4 \tag{8.77}$$

We choose the trial Hamiltonian as

$$H_0 = -\frac{1}{2}\frac{\partial^2}{\partial M^{a2}} + \frac{1}{2}\omega^2 trM^2 \tag{8.78}$$

where ω is a variational parameter.

The calculation is done in entirely the same way as in the previous section. We first note

$$<M^a(\tau)M^b(\tau')> = \delta_{ab}\,\Delta(\tau - \tau') \tag{8.79}$$

where Δ is given by (8.47) and (8.48). Then

$$<S - S_0> = ½(1 - \omega^2)\int_0^\beta d\tau <trM^2(\tau)>_0 + \frac{g}{N}\int_0^\beta d\tau <trM^4>_0$$

$$= \frac{\beta N^2}{4}[\omega^{-1} - \omega + 2g^* / \omega^2] \tag{8.80}$$

Since $W_0(\beta)$ is given by $\beta N^2\omega / 2$, we obtain

$$\frac{W_{eff}}{\beta N^2} \equiv E_0(\omega,g) = \frac{1}{4}[\omega + \omega^{-1} + 2g^* / \omega^2] \tag{8.81}$$

where g^* is given by

$$g^* = g(1 + \frac{1}{2N^2}) \tag{8.82}$$

By minimizing $E_0(\omega,g)$ we obtain a simple expression for the upper bound of the ground state energy:

$$E_0 = \frac{1}{8}(3\omega + \omega^{-1}) \tag{8.83}$$

where omega is the positive solution of

$$\omega^3 = \omega + 4g^* \tag{8.84}$$

The corresponding expression for the large N has been given in the previous chapter (7.86). We show the numerical result of (8.82) and (8.83) and compare with that of the large N.

g^*	E_0	$E_0(N=\infty)$
.01	.505	.505
.1	.5426	.542
.5	.6527	.651
1.0	.7432	.740
50.	2.236	2.217
1000.	5.968	5.915

The strong coupling result of (8.82) and (8.83) is easily obtained.

$$E_0 \approx 4^{1/3}\frac{3}{8}g^{*\,1/3} = 0.59538g^{*\,1/3} \tag{8.85}$$

which should be compared with the large N result (7.87).

We should emphasize that the variational calculation does not assume large N, so if the variational calculation is good, the result should provide a good upper bound for an arbitrary N. To see this, we first calculate E_0 for $N = 1$ in the strong coupling limit and compare the result with the exact numerical calculation of Hioe and Montroll. The variational result is

$$E_0 = 0.681428 g^{1/3} \tag{8.86}$$

and the exact numerical result is

$$E_0 = 0.66799 g^{1/3} \tag{8.87}$$

IX. WKB Method I. Instantons

9.1 Steepest Descent Method of Integration:

In the Feynman path integrals we come across integration of the quantity exp(*Action* / $\hbar$). In WKB method we first consider the limit $\hbar \to 0$ (the classical limit) and then evaluate the integral for small $\hbar$.

Let us first analyze the integration of the following type

$$Z(g) = \int_{-\infty}^{\infty} dx\, e^{-S(x)/g^2} \tag{9.1}$$

Our problem is essentially to evaluate $Z(g)$ for $g \ll 1$. The main contribution to this integral is due to the neighborhood of stable stationary points which satisfy

$$\frac{\partial S}{\partial x} = 0, \qquad S'' > 0 \tag{9.2}$$

One can approximate the function $S(x)$ by a series of quadratic forms at stationary points (minima) as shown by the dotted lines of Fig.9-1.

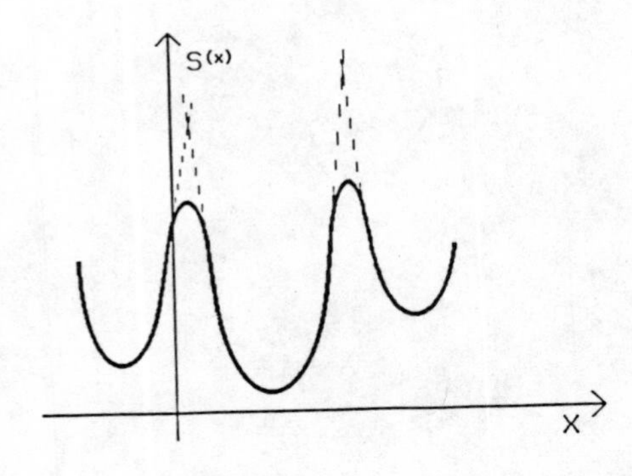

Fig.9-1

$$Z(g) = \sum_i \int dx\ e^{-S(x_i^0)/g^2} e^{-\frac{1}{2}S''(x_i^0)(x - x_i^0)^2/g^2} \tag{9.3}$$

where x_i^0 is the i-th minimum and the limits of integration are properly chosen.

Now let us take the following successive transformations

1) $x \to x + x_i^0$.

2) $x \to gx$

The limits of integration also change such that in the limit $g \to 0$, the limits are from $-\infty$ to ∞.

$$Z(g) = \sum_i e^{-S(x_i^0)/g^2} g \int_{-\infty}^{\infty} dx e^{-S''(x_i^0)x^2/2} \tag{9.4}$$

The integration is now a Gaussian integration. Hence,

$$Z(g) = \sum_i e^{-S(x_i^0)/g^2} \left| \frac{2\pi g^2}{S''(x_i^0)} \right|^{1/2} \tag{9.5}$$

This method of integration is called the method of steepest descent. This is a prototype of WKB method. We generalize this case to the Feynman path integral. The correspondence is

$S \longleftrightarrow$ *action*

$\frac{\partial S}{\partial x} = 0 \longleftrightarrow$ *classical equation of motion.*

$S'' > 0 \longleftrightarrow$ *stability of the clasical solution*

So the steepest descent method is the same as expanding the path integral variables about the classical solutions and integrating out the fluctuations by Gaussian integration. The stability of the classical solution assures that the integration of the fluctuations is a Gaussian type.

9.2 Double Well Potential: an Example.

Let us take a potential of the form

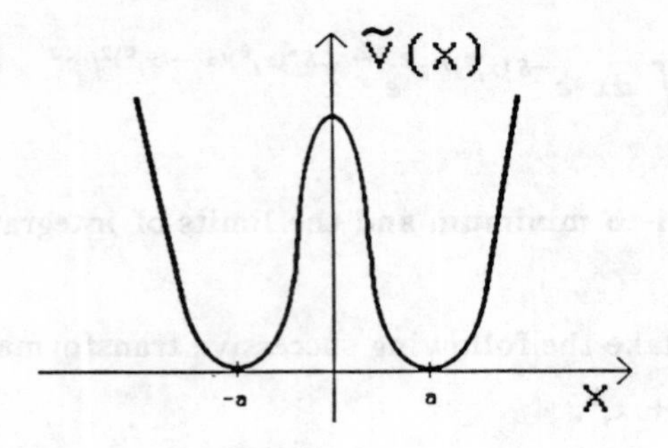

Fig.9-2

$$V(q) = \frac{1}{g^2}\tilde{V}(x), \qquad q = \frac{1}{g}x \tag{9.6}$$

where $\tilde{V}(x)$ is given by

$$\tilde{V}(x) = \frac{\omega^2}{8a^2}(x^2 - a^2)^2 \tag{9.7}$$

The form of the potential for $g \ll 1$ is obtained from $\tilde{V}(x)$ by stretching Fig.9-2 in the vertical direction by the factor $1/g^2$ and in the horizontal direction by the factor $1/g$ keeping the curvature at the bottom of the potential constant. Therefore, in the weak coupling case the two wells are separated by a high barrier and in the limit of $g=0$ there exists a pair of degenerate ground states, each of which corresponds to the ground state of harmonic potential located at $x=+a$ and $x=-a$ respectively. For a finite g there is a tunneling between the two wells resulting in the mixing of two degenerate states. The degenerate states then split into two, the one symmetric and the other anti-symmetric. The symmetric state is the ground state. In this section we calculate the ground state energy.

Let the Hamiltonian of the system be

$$\hat{H} = \frac{\hat{p}^2}{2} + V(\hat{q}) \tag{9.8}$$

then the partition function is given by

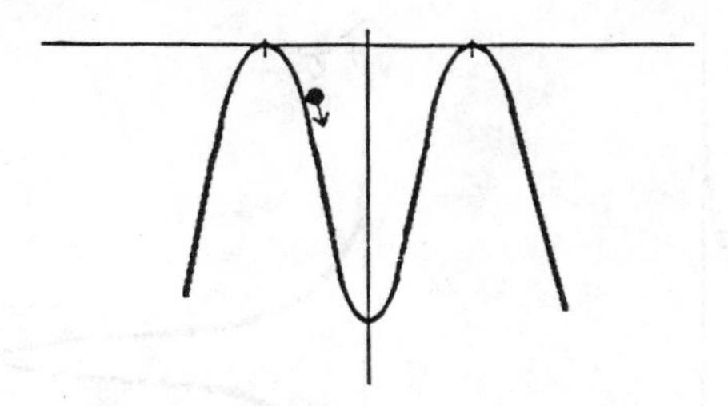

Fig.9-3

$$Z(\beta) = tr e^{-\beta \hat{H}}$$

$$= \int_{q(0)=q(\beta)} \cdots \int Dq e^{-\int_0^\beta (\frac{1}{2}\dot{q}^2 + V(q(\tau)))d\tau}$$

$$= \int_{x(0)=x(\beta)} \cdots \int \frac{Dx}{g} e^{-\frac{1}{g^2}\int_0^\beta d\tau(\frac{1}{2}\dot{x}^2 + \tilde{V}(x))} \tag{9.9}$$

We apply the method of steepest descent to this integral. Throughout this section we keep the following scale relation between $q(\tau)$ and $x(\tau)$:

$$q(\tau) = \frac{1}{g} x(\tau)$$

We need stationary points, which are solutions of

$$-\ddot{x} + \tilde{V}'(x) = 0, \qquad \tilde{V}'(x) = \frac{\partial \tilde{V}(x)}{\partial x} \tag{9.10}$$

This equation is identical to the mechanical equation of motion of a point particle in the potential $-\tilde{V}$. (See Fig.9-3.) Therefore, from the mechanical analogy all the solutions are easily found.

First, the static solutions are given by

$$x(\tau) = \pm a \tag{9.11}$$

each of which corresponds to the particle at rest on the top of the hill. It

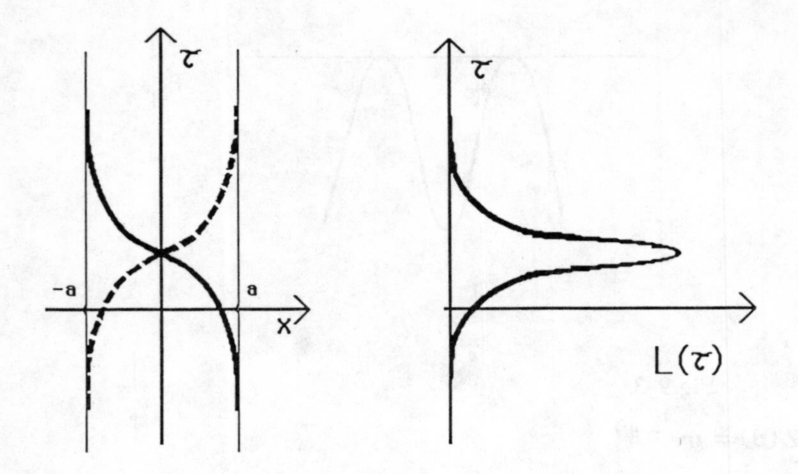

Fig. 9-4 Fig. 9-5

is an unstable situation as a mechanical motion. However, if one inserts (9.11) into the action one finds the value of the action is zero. Since the action is positive definite, solutions (9.11) correspond to the absolute minima. This indicates that the use of mechanical analogy is good only to find solutions. About the stability of the solutions we should rely on other considerations.

There exists a one parameter family of solutions of the form

$$x(\tau) = x_0(\tau - \tau_0), \qquad x_0(\tau) = a \tanh\left(\frac{\pm\tau}{\sqrt{2}}\right) \tag{9.12}$$

This solution corresponds to the motion of the particle which starts from one hill and ends at the other hill. The functional form of $x_0(\tau)$ is shown in Fig.9-4.

Each solution can be classified by τ_0 at which the particle reaches the bottom of the valley. The contribution of this solution to the action is mainly from the instantaneous kink. See Fig.9-5 where the solid bold line represents the Lagrangian density

$$L = \frac{1}{2}\dot{x}_0^2 + V(x_0(\tau))$$

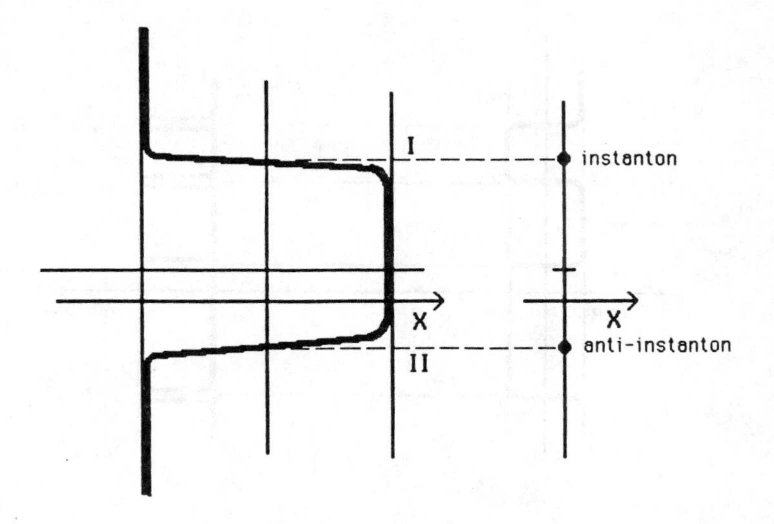

Fig 9-6

Because of this instantaneous contribution of the kink solution, we call it an instanton (or anti-instanton) located at τ_0 . An instanton and an anti-instanton occur only in pairs, because one instanton or one anti-instanton alone will change the initial and final states either from $x=+a$ to $x=-a$ or from $x=-a$ to $x=+a$, while since we are taking trace, the initial state should be the same as the final state (see Fig.9-6).

For large β , it is not difficult to see that many instanton-anti-instanton pairs also consist an approximate solution when they are well separated in time. Let us denote a solution with n pairs of instantons and anti- instantons by $x_0^{(2n)}$, which can be specified by the position of pairs as indicated by Fig. 9-7. If the instantons and anti-instantons are well separated in time then we can treat them as a one dimensional non-interacting gas. This approximation is called the dilute gas approximation.

In the dilute gas approximation we neglect the overlap of instantons and anti-instantons so that the classical action is expressed as a sum of individual instanton and anti-instanton actions.

$$S^{(2n)} = 2nS_0 \tag{9.13}$$

where S_0 is the classical action for a single instanton. We used also the

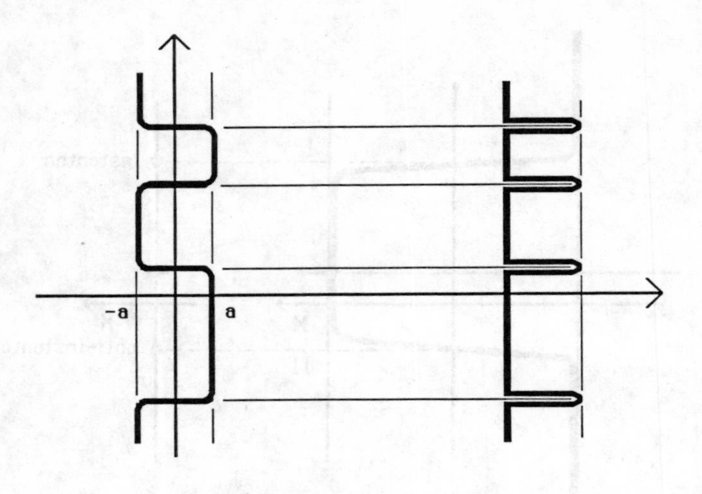

Fig.9-7

fact that the classical action of an anti-instanton is the same as that of an instanton.

Let us analyze first the $n=1$ contribution to the partition function in detail. As shown in Fig.9-6, the stationary path is the configuration of an instanton-anti-instanton pair. We evaluate the path integral by expanding the path variables about this stationary path. Since the path integral is a multiple integral we can factorize the integral into two integrals of region I and II of Fig.9-6. In I we have an instanton configuration while in II an anti-instanton. We then multiply and divide by the zero instanton contribution in region I and II:

$$Z_2 = \int \cdots \int \prod_\tau dx(\tau) e^{-\int_0^\beta d\tau L(\tau)}$$

$$= \int_{(1)} \cdots \int \prod_{\tau \in I} dx(\tau) e^{-\int_I d\tau L(\tau)} \int_{(\bar{1})} \cdots \int \prod_{\tau \in II} dx(\tau) e^{-\int_{II} d\tau L(\tau)}$$

$$= \frac{\left[\int_{(1)} \cdots \int \prod_{\tau \in I} dx(\tau) e^{-\int_I d\tau L(\tau)} \int_{(0)} \cdots \int \prod_{\tau \in II} dx(\tau) e^{-\int_{II} d\tau L(\tau)}\right] \times}{\int_{(0)} \cdots \int \prod_{I+II} dx(\tau) e^{-\int_{I+II} d\tau L(\tau)}}$$

$$\times \left[\int_{(0)} \cdots \int \prod_{\tau \in I} dx(\tau) e^{-\int_I d\tau L(\tau)} \int_{(\bar{I})} \cdots \int \prod_{\tau \in II} dx(\tau) e^{-\int_{II} d\tau L(\tau)}\right] \tag{9.14}$$

The first bracket of (9.14) is the one instanton path integral while the second bracket is the anti-instanton one. However, still there is a restriction that the position of instanton τ_0 is larger than that of anti-instanton $\bar{\tau}_0$. If we remove this restriction we have to divide by 2. So, we obtain

$$Z_2 = \frac{1}{2} Z_1^2 / Z_0 = \frac{1}{2!} R^2 Z_0 \tag{9.15}$$

where Z_1 and Z_0 are the partition functions of one and zero instanton respectively and we already used the fact that the partition function of anti-instanton is the same as Z_1 by symmetry.

$$R = \frac{Z_1}{Z_0} = \frac{\int_{(1)} \cdots \int Dx e^{-\int_0^\beta L d\tau}}{\int_{(0)} \cdots \int Dx e^{-\int_0^\beta L d\tau}} \tag{9.16}$$

The subscripts (1) and (0) with the integration sign represent the one and zero instanton configurations.

We can generalize this result to n pairs:

$$Z_{2n} = \frac{1}{(2n)!} R^{2n} Z_0 \tag{9.17}$$

So, now we have

$$Z(\beta) = Z_0 + Z_2 + Z_4 + \cdots$$

$$= Z_0 \sum_{n=1}^{\infty} \frac{1}{(2n)!} R^{2n}$$

$$= Z_0 \cosh R \underset{\beta \to \infty}{\longrightarrow} e^{-\frac{\beta}{2}\omega + R} \tag{9.18}$$

where Z_0 is the partition function of zero instanton which can be evaluated by expanding $x(\tau)$ about a and can be shown to be the same as the partition function of free harmonic oscillator:

$$Z_0 = \sum_{n=0}^{\infty} e^{-\beta(n+\frac{1}{2})\omega} \underset{\beta \to \infty}{\longrightarrow} e^{-\frac{1}{2}\beta\omega} \tag{9.19}$$

In (9.18) we used the fact that R is a positive constant proportional to β, which will be proven later (see (9.45)).

Let ϵ be the shift of energy due to the instanton phenomena (tunneling)

$$E_0 = \frac{\omega}{2} + \epsilon \tag{9.20}$$

Then we obtain

$$\epsilon = -\lim_{\beta \to \infty} \frac{1}{\beta} R \tag{9.21}$$

$$R = \frac{Z_1}{Z_0} = \frac{\underset{\text{one instanton}}{\int \cdots \int} Dq\, e^{-\int d\tau (\frac{1}{2}\dot{q}^2 + V(q))}}{\underset{\text{zero instanton}}{\int \cdots \int} Dq\, e^{-\int d\tau (\frac{1}{2}\dot{q}^2 + V(q))}} \tag{9.22}$$

Using the following expansion

$$q(\tau) = \frac{1}{g} x_0(\tau - \tau_0) + \xi(\tau - \tau_0) \tag{9.23}$$

for Z_1 and

$$q(\tau) = -\frac{a}{g} + \xi(\tau) \tag{9.24}$$

for Z_0 we obtain

$$R = e^{-S_0} \frac{\int \cdots \int D\xi\, e^{-\int_0^\beta \frac{1}{2}(\dot{\xi}^2 + U\xi^2) d\tau}}{\int \cdots \int D\xi\, e^{-\int_0^\beta \frac{1}{2}(\dot{\xi}^2 + \omega^2\xi^2) d\tau}}$$

$$= e^{-S_0} \left[\frac{\det(-\partial_\tau^2 + U(\tau))}{\det(-\partial_\tau^2 + \omega^2)} \right]^{-1/2} \tag{9.25}$$

where

$$S_0 = \int_{-\infty}^{\infty} d\tau \left(\frac{1}{2}\dot{q}_0^2 + V(q_0)\right)$$

$$= \frac{1}{g^2} \int_{-\infty}^{\infty} d\tau \left(\frac{1}{2}\dot{x}_0^2 + \tilde{V}(x_0(\tau))\right) \tag{9.26}$$

which is $O(1/g^2)$ and

$$U(\tau) = V''(q_0(\tau)) \tag{9.27}$$

The evaluation of the determinant can be done in principle by solving Schrödinger-like differential equation

$$\left(-\frac{\partial^2}{\partial\tau^2} + U(\tau)\right)\psi_n = E_n \psi_n \tag{9.28}$$

The form of $U(\tau)$ is shown in Fig.9-8.

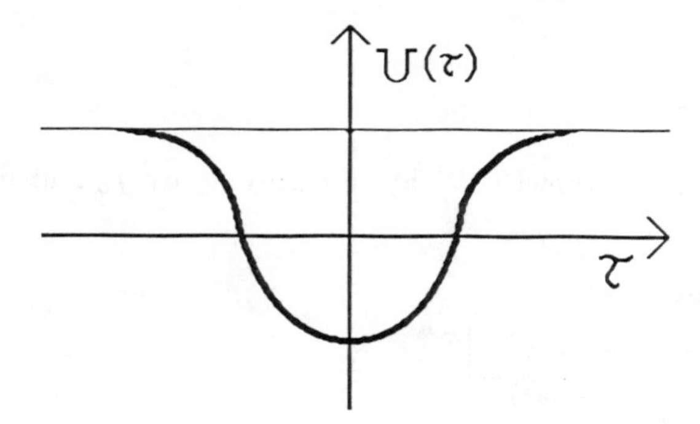

Fig.9-8

If there are solutions for zero energy modes we get an infinity for $R = Z_1/Z_0$. In the present case there is a zero energy mode and the infinity associated with it will be canceled by $\frac{1}{\beta}$ in $\frac{1}{\beta}(Z_1/Z_0)$.

Let us first demonstrate the existence of zero energy mode. $q_0(\tau)$ satisfies the classical equation

$$-\partial_\tau^2 q_0(\tau) + V'(q_0(\tau)) = 0 \tag{9.29}$$

Taking a τ derivative one obtains

$$-\partial_\tau^2 \dot{q}_0 + V''(q_0)\dot{q}_0 = 0 \tag{9.30}$$

Let

$$\dot{q}_0(\tau) \propto \psi_0(\tau) \tag{9.31}$$

then $\psi_0(\tau)$ is a zero energy solution.

We normalize ψ_0 as

$$\int d\tau \psi_0^2(\tau) = 1 \tag{9.32}$$

so that

$$\psi_0(\tau) = \dot{q}_0(\tau)/\sqrt{S_0} \tag{9.33}$$

where S_0 is the value of the classical action of an instanton, given by (9.26). In the last step of the previous equation we used

$$\frac{1}{2}\dot{q}_0^2 = V(q_0(\tau)) \tag{9.34}$$

so that

$$S_0 = \int d\tau \dot{q}_0^2$$

which can be derived from (9.29) by multiplying by $\dot{q}_0$ and integrating over τ.

Next, we prove

$$\lim_{\beta\to\infty}\frac{1}{\beta}\left[\frac{\det(-\partial_\tau^2 + U(\tau))}{\det(-\partial_\tau^2 + \omega^2)}\right]^{-1/2}$$

$$= \left(\frac{S_0}{2\pi}\right)^{1/2}\left[\frac{\det'(-\partial_\tau^2 + U(\tau))}{\det(-\partial_\tau^2 + \omega^2)}\right]^{-1/2} \tag{9.35}$$

where det' is the determinant without the zero energy mode, namely

$$\det'(-\partial_\tau^2 + U(\tau)) = \prod_{n\neq 0} E_n \tag{9.36}$$

and we normalize as

$$\frac{\det'(-\partial_\tau^2 + U(\tau))}{\det(-\partial_\tau^2 + \omega^2)} = \frac{\prod_{n \neq 0} E_n}{\prod_n E_n^{(0)}} \tag{9.37}$$

where $E_n^{(0)}$ is the energy eigenvalue of free harmonic oscillator:

$$(-\partial_\tau^2 + \omega^2)\psi_n^{(0)} = E_n^{(0)}\psi_n^{(0)}, \qquad E_n^{(0)} = (n + \frac{1}{2})\omega \tag{9.38}$$

This normalization fixes the path integration measure in (9.25) up to a factor. Namely, using the expansion

$$\xi(\tau) = \sum_n \xi_n \psi^{(0)}(\tau) \tag{9.39}$$

we obtain

$$\int \cdots \int D\xi \, e^{-\int \frac{1}{2}(\dot{\xi}^2 + \omega^2\xi^2)d\tau}$$

$$= \int \cdots \int \prod_n \frac{d\xi_n}{(2\pi)^{1/2}} e^{-\frac{1}{2}\sum_n E_n^{(0)}\xi_n^2} = \left[\prod_n E_n^{(0)}\right]^{-1/2}$$

So, we define

$$D\xi = \prod_n \frac{d\xi_n}{(2\pi)^{1/2}} \tag{9.40}$$

In order to show (9.35) we must treat the zero energy mode properly. We use the collective coordinate method for this purpose. Let us look at the numerator of (9.25):

$$\int \cdots \int D\xi \, e^{-\int_0^\beta \frac{1}{2}(\dot{\xi}^2 + U\xi^2)d\tau}$$

which was derived from the numerator of (9.22) by the change of variables (9.23). We expand ξ in terms of ψ_n

$$\xi(\tau) = \sum_n \xi_n \psi_n(\tau) \tag{9.41}$$

and change the integration variable to ξ_n. Since ψ_0 is a zero energy solution, there is no ξ_0 dependence in the exponent so that the ξ_0 integration is not of Gaussian type. We convert the ξ_0 integration to the τ_0 integration as follows. Note first,

$$\int_{-\infty}^{\infty} d\tau q(\tau+\tau_0)\psi_0(\tau) = \int_{-\infty}^{\infty} d\tau q_0(\tau)\psi_0(\tau) + \int_{-\infty}^{\infty} \xi(\tau)\psi_0(\tau) d\tau$$

$$= \frac{1}{2\sqrt{S_0}} \int_{-\infty}^{\infty} \left(\frac{\partial}{\partial\tau} q_0^2(\tau)\right) d\tau + \xi_0 = \xi_0 \tag{9.42}$$

and

$$\frac{\partial}{\partial\tau_0} \int_{-\infty}^{\infty} d\tau\, q(\tau+\tau_0)\psi_0(\tau) = \int_{-\infty}^{\infty} d\tau\, \dot{q}_0(\tau)\psi_0(\tau) + \int_{-\infty}^{\infty} \dot{\xi}(\tau)\psi_0(\tau)$$

$$= S_0^{1/2} + 0(1) \tag{9.43}$$

Thus,

$$\frac{d\xi_0}{(2\pi)^{1/2}} = (S_0/2\pi)^{1/2} d\tau_0 \tag{9.44}$$

The integration of ξ_0 is therefore done effectively by multiplying by $\beta(S_0/2\pi)^{1/2}$, accordingly we obtain (9.35).

Now,

$$R = \beta e^{-S_0}(S_0/2\pi)^{1/2} \left| \frac{\det'(-\partial_\tau^2 + U)}{\det(-\partial_\tau^2 + \omega^2)} \right|^{-1/2} \tag{9.45}$$

In order to evaluate [det'/ det] let us define

$$\Delta(E) \equiv \frac{\det(-\partial_\tau^2 + U - E)}{\det(-\partial_\tau^2 + \omega^2 - E)} = \frac{\prod_n (E_n - E)}{\prod_n (E_n^{(0)} - E)} \tag{9.46}$$

then

$$\frac{\det'(-\partial_\tau^2 + U)}{\det(-\partial_\tau^2 + \omega^2)} = \lim_{E\to 0}\left(-\frac{1}{E}\Delta(E)\right) = -\Delta'(0) \tag{9.47}$$

Let us consider a one-dimensional scattering problem

$$(-\partial_\tau^2 + U(\tau))\psi(\tau) = E\psi(\tau) \tag{9.48}$$

$\Delta(E)$ is called the Fredholm determinant associated with this scattering problem. (Note $U(\pm\infty) = \omega^2$.) As in the standard quantum mechanics book, let us define the solutions of (9.48) as $f_\pm(\tau, E)$, which satisfy the following boundary conditions

$$\lim_{\tau\to\pm\infty}[f_{\pm}(\tau,E)\,e^{ik\tau}] = 1, \qquad k^2 = E - \omega^2 \tag{9.49}$$

Let the asymptotic form of $f_{\pm}(\tau,E)$ to the opposite direction be

$$f_{\pm}(\tau,E) \underset{\tau\to\ \infty}{\to} e^{\ ik\tau} A_{\pm}(E) + e^{\pm ik\tau} F_{\pm}(E) \tag{9.50}$$

then it is known that

$$F_{+}(E) = F_{-}(E) = \Delta(E) \tag{9.51}$$

We give a proof at the end of this section.

We define a constant { K } by the asymptotic form of the instanton solution

$$q_0(\tau) \equiv \frac{1}{g}\,x_0(\tau)$$

$$\underset{\tau\to\infty}{\to} \frac{a}{g} - \frac{K}{\omega}\,e^{-\omega\tau} \tag{9.52}$$

$$\dot{q}_0(\tau) \underset{\tau\to\infty}{\to} Ke^{-\omega\tau} \tag{9.53}$$

Since $\dot{q}_0(\tau)$ is a zero energy solution of (9.48), we must have

$$f_{\pm}(\tau,0) = \dot{q}_0(\tau)/\,K \tag{9.54}$$

In fact, using (9.49) and (9.50) we obtain the asymptotic form of $f_{\pm}(\tau,0)$ as

$$f_{\pm}(\tau,0) \underset{\tau\to\pm\infty}{\to} e^{-\omega|\tau|}$$

$$f_{\pm}(\tau,0) \underset{\tau\to\ \infty}{\to} e^{-\omega|\tau|} A_{\pm}(0) + e^{\omega|\tau|}\Delta(0) \tag{9.55}$$

since at $E=0$

$$k = i\omega \tag{9.56}$$

From (9.55) we conclude that

$$A_{\pm}(0) = 1, \qquad \Delta(0) = 0 \tag{9.57}$$

The last relations are of course the expected ones.

Let us define Wronskian by

$$W[A(\tau),B(\tau)] = A(\tau)\frac{\partial}{\partial\tau}B(\tau) - B(\tau)\frac{\partial}{\partial\tau}A(\tau) \tag{9.58}$$

then it is easy to derive

$$\frac{\partial}{\partial\tau}W[f_+(\tau,E),f_-(\tau,E')] = (E-E')f_+(\tau,E')f_-(\tau,E') \tag{9.59}$$

Thus,

$$\frac{\partial^2}{\partial E\,\partial\tau}W[f_+(\tau,E),f_-(\tau,0)]\Big|_{E=0} = f_+(\tau,0)f_-(\tau,0) = \frac{\dot{q}_0^2(\tau)}{K^2} \tag{9.60}$$

Integrating this over τ from $-\infty$ to $+\infty$ we obtain

$$\lim_{E\to 0}\left[\frac{\partial}{\partial E}W[f_+(\tau,E),f_-(\tau,0)]\right]_{-\infty}^{\infty} = \frac{1}{K^2}S_0 \tag{9.61}$$

We compute the left hand side of (9.61) by using the asymptotic form of $f_\pm(\tau,E)$ (9.49) and (9.50). Then we obtain $-2\omega\Delta'(0)$ so that

$$\Delta'(0) = -\frac{S_0}{2\omega K^2} \tag{9.62}$$

Thus,

$$\begin{aligned}\epsilon &= -\lim_{\beta\to\infty}\frac{1}{\beta}R\\ &= e^{-S_0}(S_0/2\pi)^{1/2}\left|\frac{\det'(-\partial_\tau^2+U)}{\det(-\partial_\tau^2+\omega^2)}\right|^{-1/2}\\ &= e^{-S_0}(S_0/2\pi)^{1/2}\left(-\frac{1}{\Delta'(0)}\right)^{1/2}\\ &= K(\omega/\pi)^{1/2}e^{-S_0}\end{aligned} \tag{9.63}$$

Proof of (9.51):

By setting $E=E'$ in (9.59) we see that $W[f_+(\tau,E),f_-(\tau,E)]$ is independent of τ. So we compute it at $\tau=\pm\infty$ and we obtain

$$W[f_+(\tau,E),f_-(\tau,E)] = -2ikF_+(E) = -2ikF_-(E) \tag{9.64}$$

which proves the first part of (9.51). From now on we simply write $F(E)$ for $F_\pm(E)$.

Next we note

$$G(\tau,\tau';E) = \frac{if_+(\tau_>,E)f_-(\tau_<,E)}{2kF(E)} \tag{9.65}$$

is a Green's function of $-\partial_\tau^2 + U(\tau)$:

$$\left[-\frac{\partial^2}{\partial\tau^2} + U(\tau) - E\right]G(\tau,\tau';E) = \delta(\tau-\tau') \tag{9.66}$$

The symbols $\tau_>$ and $\tau_<$ are defined by

$$\tau_> = \tau, \qquad \tau_< = \tau' \quad \text{if} \quad \tau > \tau'$$

$$\tau_> = \tau', \qquad \tau_< = \tau \quad \text{if} \quad \tau < \tau' \tag{9.67}$$

We may write

$$G(\tau,\tau';E) = \langle\tau|(\hat{H}-E)^{-1}|\tau'\rangle \tag{9.68}$$

where

$$\langle\tau|\hat{H}|\tau'\rangle = \left[-\frac{\partial^2}{\partial\tau^2} + U(\tau)\right]\delta(\tau-\tau') \tag{9.69}$$

With this notation

$$\Delta(E) = \frac{\det(\hat{H}-E)}{\det(\hat{H}_0-E)} \tag{9.70}$$

So

$$\frac{\partial}{\partial E}\ln\Delta(E) = -tr\left[\frac{1}{\hat{H}-E} - \frac{1}{\hat{H}_0-E}\right]$$

$$= -\int d\tau[G(\tau,\tau;E) - G_0(\tau,\tau;E)] \tag{9.71}$$

Using (9.59), we obtain

$$f_+(\tau,E)f_-(\tau,E) = \lim_{E'\to E}\frac{1}{E-E'}\frac{\partial}{\partial\tau}W[f_+(\tau,E),f_-(\tau,E')]$$

So we can compute (9.71) by

$$\frac{\partial}{\partial E}\ln\Delta(E) = -\lim_{T\to\infty}\lim_{E'\to E}\frac{i}{E-E'}\left[\frac{W[f_{+}(\tau,E),f_{-}(\tau,E')]}{2kF(E)} - \right.$$

$$\left. - \frac{W[e^{ik\tau},e^{-ik\tau}]}{2k}\right|_{-T}^{T}$$

$$= \frac{F'(E)}{F(E)} = \frac{\partial}{\partial E}\ln F(E) \qquad (9.72)$$

Since

$$\Delta(\infty) = 1$$

by definition (9.70) and

$$F(\infty) = 1$$

which can be easily understood because at high energy the solution approaches the free solution, we obtain

$$\Delta(E) = F(E) \qquad (9.73)$$

(QED)

9.3 Ground State Energy of Double Well Potential in Terms of the Standard WKB Calculation

In the previous section we obtained the ground state energy of double well potential by using the instanton method, which is a typical WKB method in path integral. The purpose of this section is two-fold; a) to show that the instantons are relevant to the familiar tunneling phenomena in quantum mechanics and b) to prepare for the WKB method for many variable systems which we shall develop in **XI** .

In the standard WKB method we calculate the logarithm of wave function in the power series of $\hbar$ in various regions of configuration space and then connect them in a common asymptotic region. As an example we discuss the double well potential problem. In this case we first calculate the wave function in the forbidden region, which is the region between the two wells. Next we calculate the wave function in the allowed region (i.e. $q\approx a$). Then connect them in a common region which we assume exists.

The Schrödinger equation of the system is given by

$$\left[-\frac{1}{2}\frac{\partial^2}{\partial q^2} + V(q)\right]\psi = E\psi \tag{9.74}$$

Scaling by g $(q \equiv x/g)$

$$\frac{1}{g^2}\left[-\frac{g^4}{2}\frac{\partial^2}{\partial x^2} + \tilde{V}(x)\right]\psi = E\psi \tag{9.75}$$

where $\tilde{V}(x)$ is given by (9.7) and is independent of g.

Let us first calculate the wave function in the forbidden region. As usual

$$\psi = e^S, \qquad S = \frac{1}{g^2}W_0 + W_1 + \cdots \tag{9.76}$$

$$E = \frac{1}{g^2}\epsilon_0 + \epsilon_1 + \cdots \tag{9.77}$$

insert them into the Schrödinger equation and equate terms of the order of g^2. We obtain

$$-\frac{1}{2}(W_0')^2 + \tilde{V}(x) = \epsilon_0 \tag{9.78}$$

$$-\frac{1}{2}W_0'' - W_0'W_1' = \epsilon_1 \qquad \textit{etc.} \tag{9.79}$$

The first equation is the Hamilton-Jacobi equation so that the solution can be obtained in terms of the classical solution. We use the instanton solution discussed in the previous section. Namely, for $\epsilon_0 = 0$ by comparing (9.78) with (9.34) we set

$$W_0'(x) = \pm\dot{x}_0(\tau) \tag{9.80}$$

where the relation between x and τ is

$$x = x_0(\tau)$$

or equivalently

$$q = q_0(\tau) \tag{9.81}$$

Multiplying (9.80) by $\dot{x}_0$ and integrating over τ we obtain

$$\frac{1}{g^2}W_0(x) = \pm\int_0^{\tau} \dot{q}_0^2(\tau')\, + const \tag{9.82}$$

The solution of W_1 is given by

$$W_1(x) = -\frac{1}{2}\ln|W_0^2(x)| - \epsilon_1 \int^{x=gq} \frac{dx'}{W_0(x')} + const$$

$$= -\ln|\dot{q}_0(\tau)|^{1/2} - \epsilon_1\tau + const \qquad (9.83)$$

Inserting (9.82) and (9.83) into (9.76) we obtain two independent WKB wave functions

$$\psi_{WKB}^{(\pm)}(q) = \frac{C}{|\dot{q}_0(\tau)|^{1/2}} e^{\pm(\int_0^{\tau} \dot{q}_0^2(\tau')d\tau' - E\tau)} \qquad (9.84)$$

where we equated $E = \epsilon_1$, since $\epsilon_0 = 0$.

Since the exponent of (9.84) is an odd function of q (note the relation of q and τ given by (9.81) and $q_0(-\tau) = -q_0(\tau)$), we obtain

$$\psi_{WKB}^{(+)}(q) = \psi_{WKB}^{(-)}(-q) \qquad (9.85)$$

Since we are interested in the ground state, we demand the wave function to be symmetric.

$$\psi_{WKB} = \psi_{WKB}^{(+)} + \psi_{WKB}^{(-)} \qquad (9.86)$$

For large τ, i.e. $q \sim a$ we use (9.52) so that

$$\int_0^{\tau} \dot{q}(\tau')d\tau' \equiv \int_0^{\infty} \dot{q}_0^2 d\tau - \int_{\tau}^{\infty} \dot{q}_0^2(\tau')d\tau'$$

$$= \frac{1}{2}S_0 - \frac{1}{2}\omega y^2 \qquad (9.87)$$

where S_0 is the classical action and y is given by

$$y = \frac{a}{g} - q \qquad (9.88)$$

As before we set

$$E = E_0 = \frac{1}{2}\omega + \epsilon \tag{9.20}$$

$\frac{1}{2}\omega$ is the ground state energy of the single well which should be approached by E in the limit of $g \to 0$. ϵ is the correction due to the tunneling effect.

Using (9.84), (9.87) and (9.53) we obtain

$$\psi_{WKB}^{(+)} = \frac{C}{K^{1/2}} e^{\frac{1}{2}S_0} e^{-\frac{1}{2}\omega y^2} \tag{9.89}$$

$$\psi_{WKB}^{(-)} = \frac{CK^{1/2}}{y} e^{-\frac{1}{2}S_0} e^{\frac{1}{2}\omega y^2} \tag{9.90}$$

This is the behavior of the wave function approaching $q \sim \frac{a}{g}$ from the forbidden region.

Near $q \sim \frac{a}{g}$ the Schrödinger wave function should be a solution of the equation with harmonic oscillator potential:

$$\left[-\frac{1}{2}\frac{\partial^2}{\partial y^2} + \frac{1}{2}\omega^2 y^2\right]\psi = E\psi \tag{9.91}$$

We expect ψ to be close to the ground state wave function ψ_0 of the harmonic oscillator so we set

$$\psi = \psi_0 + \psi_1 \tag{9.92}$$

and assume ψ_1 is of order ϵ . Inserting (9.92) and (9.20) into (9.91) we obtain

$$\left[-\frac{1}{2}\frac{\partial^2}{\partial y^2} + \frac{1}{2}\omega^2 y^2 - \frac{1}{2}\omega\right]\psi_0 = 0 \tag{9.93}$$

$$\left[-\frac{1}{2}\frac{\partial^2}{\partial y^2} + \frac{1}{2}\omega^2 y^2 - \frac{1}{2}\omega\right]\psi_1 = \epsilon\psi_0 \tag{9.94}$$

ψ_0 is of course the ground state function of the harmonic oscillator ($\psi_0 \sim e^{-\frac{1}{2}\omega y^2}$) which matches $\psi_{WKB}^{(+)}$. (Notice $\psi_{WKB}^{(+)}$ has precisely the Gaussian form.) Accordingly ψ_1 should match with $\psi_{WKB}^{(-)}$.

Using (9.93) and (9.94) we obtain the expression for the energy shift by the following Wronskian

$$\epsilon = -\frac{\psi_0 \frac{\partial}{\partial y}\psi_1 - \psi_1 \frac{\partial}{\partial y}\psi_0}{\int_{-\infty}^{y} (\psi_0(y'))^2 dy'}$$

$$\sim -\frac{\psi_{WKB}^{(+)} \frac{\partial}{\partial y}\psi_{WKB}^{(-)} - \psi_{WKB}^{(-)} \frac{\partial}{\partial y}\psi_{WKB}^{(+)}}{\int_{-\infty}^{\infty} (\psi_{WKB}^{(+)})^2 dy} \tag{9.95}$$

Since the matching region is the region

$$\omega y^2 \gg 1 \tag{9.96}$$

we extend the integration limit in the denominator to ∞. We evaluate (9.95) using (9.89) and (9.90) to obtain

$$\epsilon = K\,(\omega/\pi)^{1/2}\, e^{-S_0} \tag{9.97}$$

which agrees with (9.63).

X. WKB Method II. Solitons

We continue the discussion of the WKB method in many variable systems. In this chapter we discuss in detail the ϕ^4-scalar field theory model in 1+1 dimension. We intend to discuss the treatment of solitons in quantum field theory in terms of the semi-classical WKB method. Mathematically, as in the previous chapter we try to evaluate the Feynman path integral by expanding the paths about the stationary path. In this case, however, the stationary path is a solution of the classical equation of the theory. So we first discuss the classical solutions of scalar field theories in 1+1 dimensions.

10.1 Non-linear scalar field theory model in 2 dimensions and its classical solutions

Let us consider a two dimensional real scalar field theory model defined by the following Lagrangian density:

$$L = \frac{1}{2}(\dot{\phi}^2 - \phi'^2) - V(\phi) \tag{10.1}$$

where $\dot{\phi} = \partial\phi/\partial t$, $\phi' = \partial\phi/\partial x$. V is a function of ϕ, but as in the previous chapter the dependence of the coupling constant g is given by

$$V(\phi) = \frac{1}{g^2}\tilde{V}(X), \qquad X = g\phi \tag{10.2}$$

where $\tilde{V}$ is an even function and it is independent of g. Depending on the choice of the functional form such as Fig. 10–1 or Fig. 10–2, one can consider various models.

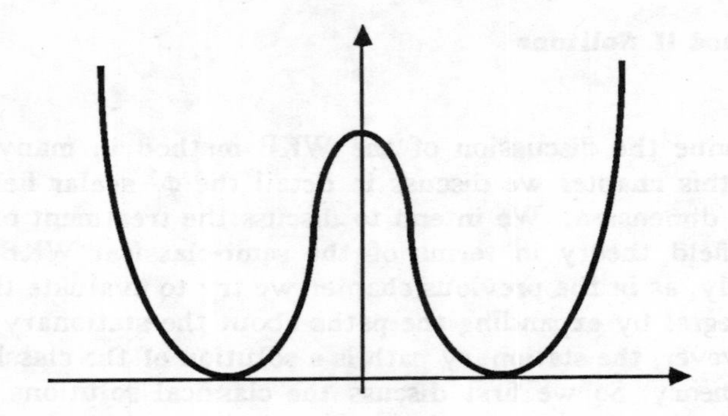

Fig. 10–1

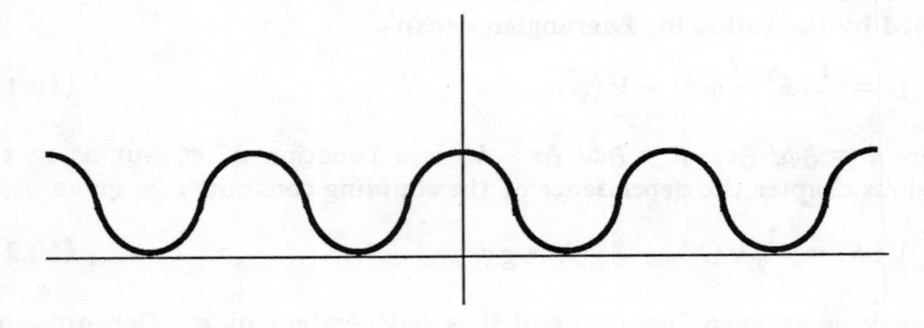

Fig. 10–2

Mechanical Analogue Model:

In order to visualize the classical solutions of the system, we make a discrete model from which the above scalar field model is obtained in the continuum limit. Let us consider a set of points equally spaced. Let x^i be the i-th point and define q^i by

$$\phi(x^i) = \frac{1}{\sqrt{\Delta x}} q^i \tag{10.3}$$

where Δx is the distance between two neighboring points. The discrete model is defined by

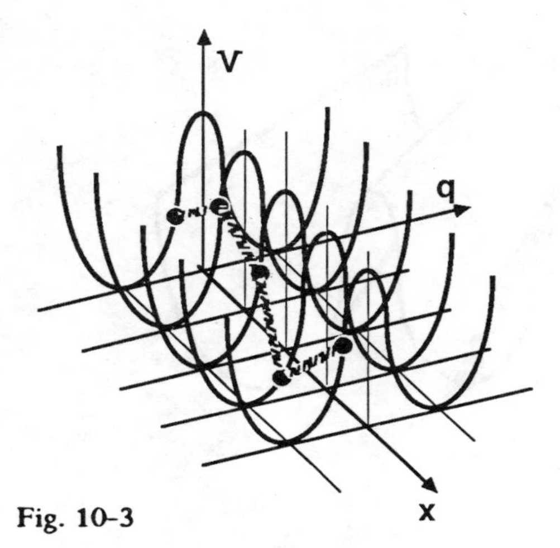

Fig. 10-3

$$L = \sum_i \left[\frac{1}{2}\dot{q}^{i2} - v(q^i) - \frac{k}{2}(q^i - q^{i-1})^2 \right] \tag{10.4}$$

The first term is the kinetic energy term which corresponds to $\dot{\phi}^2$, and the second term is the potential term.

$$v(q^i) = \frac{\Delta x}{g^2}\tilde{V}(\frac{g}{\sqrt{\Delta x}}q^i) \tag{10.5}$$

The third term corresponds to ϕ'^2 term and it represents the interaction between two points by a spring with the spring constant $k = (\Delta x)^{-2}$. Thus, the model is equivalent to a string of beads connected by springs (rosary) and placed in a uniform potential. (See Fig. 10-3.)

Classical Solutions:

The equation of motion derived from the Lagrangian (10.1) is

$$\ddot{\phi}(x,t) - \phi''(x,t) + V'(\phi(x,t)) = 0 \tag{10.6}$$

where we assume that the potential has two minima as shown in Fig. 10-3.

The lowest energy state is the case that the rosary is placed in the bottom of one of the valleys. See Fig. 10-4.

This corresponds to

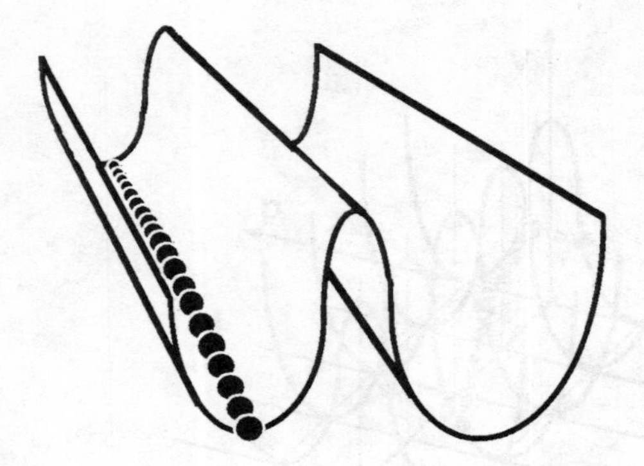

Fig. 10–4

$$\phi_c = \pm\frac{1}{g} \tag{10.7}$$

Since the classical energies of these solutions are 0, they are the lowest energy states (classical vacua). To move from one vacuum to the other we must lift the rosary over the hill. Since the length of the rosary is infinite and the height of the hill is proportional to $1/g$, it requires an infinite amount of energy.

Next let us consider the case that the one end of the rosary is placed in one of the valleys and the other end in the other valley. See Fig. 10–5.

In this configuration the energy is concentrated in the region the rosary passes the hill, since the springs are stretched there and the beads get

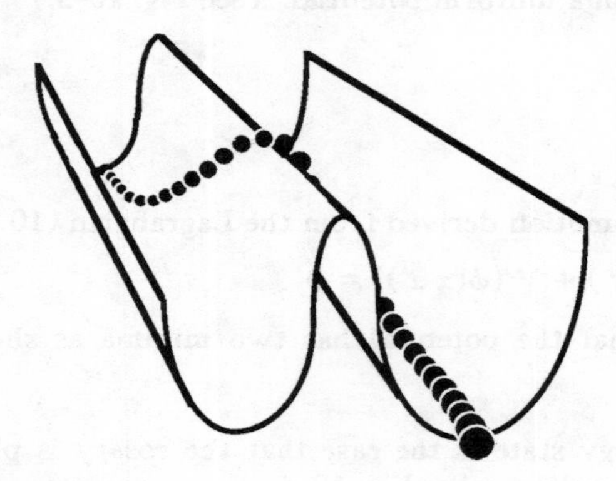

Fig. 10–5

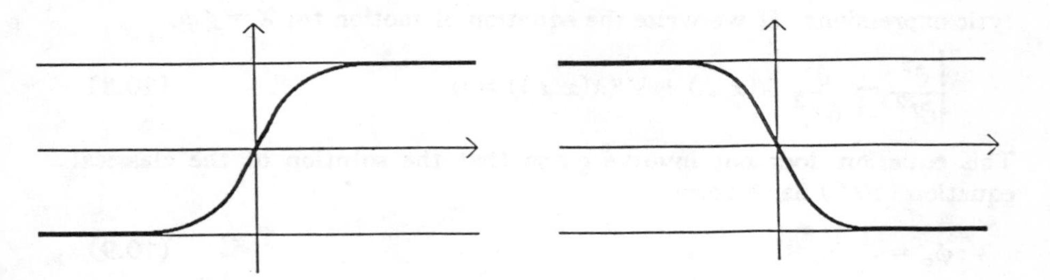

Fig. 10–6a Fig. 10–6b

potential energy. This solution corresponds to a kink solution described in Fig. 10–6a or in Fig. 10–6b.

These figures are the overview of the rosary placed in the potential well of Fig. 10-3. Fig. 10–7 is the energy distribution of these solutions.

If one pulls back a bead on the hill towards a valley, the next bead moves up and in effect the kink position moves. Thus, it is easy to imagine a moving kink solution. In this solution, since a lump of energy moves as if an extended particle, we call this kink a soliton. From the mechanical analogue model it is quite clear that this soliton solution is stable.

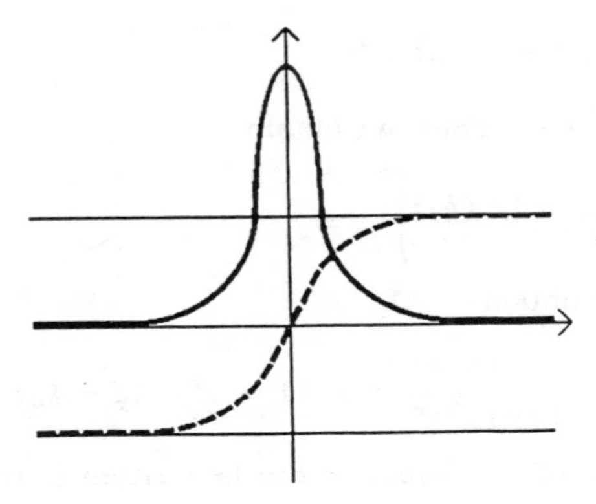

Fig. 10–7

Let us formulate the above qualitative discussions in terms of analytic expressions. If we write the equation of motion for $\chi = g\phi$.

$$\left(\frac{\partial^2}{\partial t^2} - \frac{\partial^2}{\partial x^2}\right)\chi(x\,t) + \tilde{V}'(\chi(x\,t)) = 0 \tag{10.8}$$

This equation does not involve g. so that the solution of the classical equation (10.6) has a form

$$\phi_c = \frac{1}{g}\chi_0 \tag{10.9}$$

where χ_0 is a solution of (10.8) which is independent of g. Although $\chi(x\,t)$ is a function of two variables, we first make an ansatz that the solution is a function of ξ defined by

$$\xi = x + \alpha t \tag{10.10}$$

Setting

$$\chi(x\,t) = \chi_0(x + \alpha t) \equiv \chi_0(\xi) \tag{10.11}$$

we insert it into (10.8). We obtain

$$-(1-\alpha^2)\partial_\xi^2\chi_0 + \tilde{V}'(\chi_0) = 0 \tag{10.12}$$

For large x or t, χ_0 should be at the bottom of a valley. So we impose the following boundary conditions:

$$\lim_{|\xi|\to\infty} \chi_0(\xi) = \pm 1 \tag{10.13}$$

By multiplying (10.12) by χ' and integrating over ξ we obtain

$$-\frac{1}{2}(1-\alpha^2)\chi_0'^2 + \tilde{V}(\chi_0) = 0 \tag{10.14}$$

where we used (10.13). Thus, we obtain

$$\chi_0' = \pm\left[\frac{2}{(1-\alpha^2)}\tilde{V}(\chi_0(\xi))\right]^{1/2} \tag{10.15}$$

Integrating this we obtain

$$\Phi(\chi_0(\xi)) \equiv \int^{\chi_0(\xi)} \frac{d\chi}{(2V(\chi))^{1/2}} = \pm(1-\alpha^2)^{-1}(\xi - \xi_0) \tag{10.16}$$

The right hand side of this equation can be written in terms of the original variables as

$$= \pm \frac{x - x_0 - vt}{\sqrt{1-v^2}} \tag{10.17}$$

where we set $\alpha = v$. Thus, we obtain

$$\phi_c \equiv X_0 / g = \frac{1}{g} \Phi^{-1} \left[\pm \frac{x - x_0 - vt}{\sqrt{1-v^2}} \right] \tag{10.18}$$

As we will see v corresponds to the velocity of the soliton.

ϕ^4-model:

$$\tilde{V}(X) = \frac{1}{8}(X^2 - 1)^2 \tag{10.19}$$

See Fig. 10–1.

$$\phi_c^{(\pm)} = \frac{1}{g} \tanh \left[\pm \frac{1}{2} \frac{x - x_0 - vt}{\sqrt{1-v^2}} \right] \tag{10.20}$$

x_0 is the position of the kink and v is the velocity of the kink.

Sine–Gordon model:

$$\tilde{V}(X) = 1 - \cos X \tag{10.21}$$

See Fig. 10–2

$$\phi_c^{(\pm)} = \frac{1}{4g} \tan^{-1} e^{\pm \frac{x - x_0 - vt}{\sqrt{1-v^2}}} \tag{10.22}$$

As we see in these examples there exist two one kink solutions, $\phi_c^{(+)}$ and $\phi_c^{(-)}$. + corresponds to Fig. 10–6a, while - to Fig. 10–6b. We call $\phi_c^{(+)}$ soliton and $\phi_c^{(-)}$ anti-soliton. The boundary conditions of these solutions are different from each other.

Next we obtain the energy of the soliton. The Hamiltonian density derived from the Lagrangian (10.1) is given by

$$H = \frac{1}{2}(\dot{\phi}^2 + \phi'^2) + V(\phi) \tag{10.23}$$

Thus the classical mass of the soliton is given by

$$M_0 = \int dx \left[\frac{1}{2}\phi_0'^2 + V(\phi_0)\right] \tag{10.24}$$

where ϕ_0 is the classical soliton solution at rest ($v = 0$) and it satisfies

$$-\phi_0'' + V'(\phi_0) = 0 \tag{10.25}$$

Multiplying this equation by ϕ' and integrating over x from $-\infty$ to x we obtain

$$\frac{1}{2}\phi_0'^2 = V(\phi_0) \tag{10.26}$$

Using this we obtain

$$M_0 = \int dx\, \phi_0'^2(x) \tag{10.27}$$

Since ϕ_0 is proportional to g^{-1} (see (10.18)) . the classical mass M_0 is proportional to g^{-2}. Using ϕ_0 we write ϕ_c as

$$\phi_c(x,t) = \phi_0\left(\frac{x-x_0-vt}{\sqrt{1-v^2}}\right) \tag{10.28}$$

Inserting this into (10.23) we obtain the following expression for the soliton classical energy.

$$E_0 = \frac{M_0}{\sqrt{1-v^2}} \tag{10.29}$$

Stability of the classical soliton solution:

It is quite clear intuitively from the mechanical analogue model that the soliton solution is a stable solution. Here we discuss about how this stability should be expressed mathematically.

Let us begin with the classical equation (10.8). Let χ_0 be a static soliton solution. Setting

$$\chi(x,t) = \chi_0(x) + \delta\chi(x,t) \tag{10.30}$$

we insert it into (10.8) and expand the potential term. Neglecting the higher order terms in $\delta\chi$ we obtain

$$\left(\frac{\partial^2}{\partial t^2} - \frac{\partial^2}{\partial x^2} + U(x)\right)\delta\chi(x,t) = 0 \tag{10.31}$$

where

$$U(x) = \tilde{V}''(\chi_0(x)) = V''(\phi_0(x)) \tag{10.32}$$

(10.31) is the equation for the small fluctuation $\delta\chi$.

As usual we set

$$\delta X(x,t) = e^{i\omega t}\psi(x) \tag{10.33}$$

and insert it into (10.31). We obtain

$$\left[-\frac{\partial^2}{\partial x^2} + U(x)\right]\psi(x) = \omega^2\psi(x) \tag{10.34}$$

This equation has the same form as the Schrödinger equation for an one dimensional potential problem. Let ψ_n be an eigenfunction associated with the eigenvalue ω_n^2. If eigenvalue ω_n^2 is positive δX is oscillatory in time and the magnitude of fluctuation is bounded, while if ω_n^2 is negative the solution blows up and is unstable. Thus, the stability condition is given by

$$\omega_n^2 \geqslant 0 \qquad \text{for } all\ n \tag{10.35}$$

As in the previous chapter we have a zero energy mode. Since the classical equation is translationally invariant if $X_0(x)$ is a solution so is $X_0(x+\delta x)$.

$$-X_0''(x+\delta x) + \tilde{V}'(X_0(x+\delta x)) = 0$$

Inserting $X_0(x+\delta x) \approx X_0(x) + X_0'(x)\delta x$ into the above equation and keeping the linear term in δx we obtain

$$-X_0''' + \tilde{V}''(X_0(x))X_0'(x) = 0. \tag{10.36}$$

So,

$$\psi_0(x) \propto X_0'(x) \tag{10.37}$$

In general, when the system is invariant by a continuous transformation there exists a zero mode. The zero mode in the previous chapter is associated with the time translation.

Since we know from the mechanical analogue model that the soliton solution is a stable solution, we assume the stability condition (10.35) without a proof throughout this chapter.

10.2 Perturbation Theory and Renormalization

We use a path integral representation for the quantum field theory model (10.1).

$$<F\,|\,e^{-i\hat{H}(t_f - t_i)}\,|\,I> =$$

$$= \int \cdots \int D\phi\ \psi_F^*[\phi(\cdot,t_f)]\psi_I[\phi(\cdot,t_i)]e^{i\int\int_\Omega dxdtL} \tag{10.38}$$

where Lagrangian L is given by (10.1) and Ω denotes the region of space-time;

$$\Omega: \qquad L/2 \geqslant x \geqslant -L/2, \qquad t_f \geqslant t \geqslant t_i \tag{10.39}$$

By using (10.2) the action can be written as

$$S = \frac{1}{g^2}\int\int_\Omega dxdt\,[\frac{1}{2}(\dot{X}^2 - X'^2) - \tilde{V}(X(x,t))] \tag{10.40}$$

Thus, in the weak coupling expansion we use again the steepest descent method (in this case the stationary phase method).

In the previous section we obtained classical solutions; a pair of constant solutions (classical vacua) and a pair of space dependent solutions (soliton and anti-soliton). In this section we discuss a systematic weak coupling expansion about the classical vacuum (standard perturbation expansion). As a specific model to discuss this problem we choose the ϕ^4 - model:

$$V(\phi) = \frac{m^2}{8g^2}(X^2 - 1)^2$$

$$= \frac{m^2g^2}{8}(\phi^2 - \frac{1}{g^2})^2 \tag{10.41}$$

Since $\phi_c = \frac{1}{g}$ is a classical vacuum solution, we expand the Lagrangian by setting

$$\phi(x,t) = \frac{1}{g} + \eta(x,t) \tag{10.42}$$

We obtain

$$L = \frac{1}{2}(\dot{\eta}^2 - \eta'^2 - m^2\eta^2) - \frac{m^2g}{2}\eta^3 - \frac{m^2g^2}{8}\eta^4 \tag{10.43}$$

Based on this Lagrangian one can develop a pe

Fig. 10–8a Fig. 10–8b

Since this is a field theory in 2 dimensions, the only divergent diagrams are those given by Fig. 10–8. Using the Feynman rules we evaluate these graphs and we obtain

$$\textit{Fig. } 10.8a \qquad m^2 g \frac{3}{2} \Delta \tag{10.44}$$

$$\textit{Fig. } 10.8b \qquad m^2 g^2 \frac{3}{4} \Delta \tag{10.45}$$

where

$$\Delta \equiv \Delta_F(0) = \frac{1}{(2\pi)^2} \int \frac{d^2k}{k^2 - m^2} \tag{10.46}$$

which is logarithmically divergent. In order to cancel these divergencies we add the following counter term to (10.1):

$$\delta L = -\frac{1}{4}\delta m^2(\eta^2 + \frac{2\eta}{g}) = -\frac{1}{4}\delta m^2(\phi^2 - \frac{1}{g^2}) \tag{10.47}$$

where

$$\delta m^2 = -3\,\Delta\, m^2 g^2 \tag{10.48}$$

It is easy to see that up to g^2 order this counter term is due to the renormalization of the parameters (m^2 and g^2) in the theory:

$$V(\phi) - \delta L = \frac{m^2 g^2}{8}(\phi^2 - \frac{1}{g^2})^2 + \frac{1}{4}\delta m^2(\phi^2 - \frac{1}{g^2})$$

$$= \frac{m_0^2 g_0^2}{8}(\phi^2 - \frac{1}{g_0^2})^2 + 0(g^4) \tag{10.49}$$

where

$$m_0^2 = m^2 + \delta m^2 = m^2(1 - 3\,\Delta\, g^2)$$

$$g_0^2 = g^2(1 - \frac{\delta m^2}{m^2}) = g^2(1 + 3\,\Delta\, g^2) \tag{10.50}$$

10.3 Solitons in Quantum Field Theory

Soliton Solution:

In **10.1** we discussed the classical solutions; the vacuum solution and the one soliton solution. One of the differences between these solutions is that the total momentum of the soliton solution is in general finite while it is zero for the vacuum solution. Indeed if we insert the soliton solution (10.18) into the expression for the total momentum of the system (see (10.59))

$$p = -\int dx\, \dot{\phi}(x,t)\phi'(x,t) \tag{10.51}$$

we obtain

$$p = \frac{v}{\sqrt{1 - v^2}} M_0 \tag{10.52}$$

Thus, if we impose the condition that the total momentum of the system is finite we automatically exclude the classical vacuum as a stationary phase solution.

We proceed the discussion in the phase space path integral formalism. The action is given by

$$S = \int\int dxdt\,[\pi(x,t)\dot{\phi}(x,t) - H] \tag{10.53}$$

where H is the Hamiltonian density.

$$H = \frac{1}{2}(\pi^2 + \phi'^2) + V(\phi) \tag{10.54}$$

The total momentum of the system is given by

$$P = -\int dx\, \pi(x,t)\phi'(x,t) \tag{10.55}$$

It should be an infinitesimal generator of spatial translations. Indeed one can check

$$\phi(x + \delta x) = \phi(x) + \delta\phi(x) = \phi(x) + \phi'(x)\delta x \tag{10.56}$$

$$\delta\phi(x) = [\delta x P, \phi(x)]_P = \delta x [P, \phi(x)]_P = \phi'(x)\delta x \qquad (10.57)$$

where $[.]_P$ is the Poisson bracket defined by

$$[A, B]_P = \int dx \left(\frac{\delta A}{\delta\phi(x)} \frac{\delta B}{\delta\pi(x)} - \frac{\delta B}{\delta\phi(x)} \frac{\delta A}{\delta\pi(x)} \right) \qquad (10.58)$$

We minimize the action with a subsidiary condition that the total momentum of the system is given by p:

$$p = P[\pi, \phi; t] \equiv - \int dx\, \pi(x, t)\phi'(x, t) \qquad (10.59)$$

We obtain

$$\dot{\pi} = -\phi'' + V'(\phi) - \lambda\pi' \qquad (10.60)$$

$$\dot{\phi} = \pi + \lambda\phi' \qquad (10.61)$$

where λ is a Lagrangian multiplier. We seek a static solution $(\dot{\pi} = \dot{\phi} = 0)$. From (10.60) and (10.61) we obtain

$$-(1 - \lambda^2)\phi'' + V'(\phi) = 0 \qquad (10.62)$$

This is the same equation as (10.12), $(\lambda = \alpha = v)$. So, we obtain the soliton solution as a stationary phase of the action (10.53) with the constraint (10.59).

Collective Coordinates:

We start with the following phase space path integral representation:

$$< F, t_f \mid I, t_i >$$

$$= < F \mid e^{-i\hat{H}(t_f - t_i)} \mid I > =$$

$$= \int \cdots \int D\pi D\phi \Psi_F^* \Psi_I e^{iS} \qquad (10.63)$$

$$S = \int_{t_i}^{t_f} dt \int_{-L/2}^{L/2} dx\, \pi(x, t)\dot{\phi}(x, t) - H + \delta L + \Delta E \int_{t_i}^{t_f} dt \qquad (10.64)$$

where δL is the counter term (10.47) and ΔE is an additional constant term to be adjusted such that the total energy of the vacuum becomes zero. We omit these terms until the end of the section where we will discuss the renormalization problem.

We use the following collective coordinate method based on the Faddeev-Popov technique to extract the center of mass coordinate $q(t)$ and the momentum coordinate $p(t)$. We first insert

$$\int \cdots \int \prod_t dp(t)\delta(p(t) - P[\pi,\phi;t]) = 1 \tag{10.65}$$

into the path integral to insure the total momentum of the system is given by $p(t)$. Since P in (10.65) is a generator of translations and since the center of mass coordinate $q(t)$ should be the canonical conjugate of $p(t)$, we may extract $q(t)$ through translations as follows. We insert

$$\int \cdots \int \prod_t dq(t)\delta(Q[\pi(x + q(t),t),\phi(x + q(t),t)]\frac{\partial Q}{\partial q(t)} = 1 \tag{10.66}$$

into the path integral. Q is a functional of π and ϕ in general but in practice we use the following form

$$Q[\phi(x + q(t),t)] = \int dx f(x)\phi(x + q(t),t) \tag{10.67}$$

where $f(x)$ is a function to be determined later.

In (10.65) and (10.66) we used a symbol of continuous product $\prod_t$, but the insertion of these expressions into the path integral means the following. As we explained in II , the phase space path integral is defined by the $N = \infty$ limit of a multiple integral of N p-integrations and $N-1$ q- integrations (in this case $N\ \pi(\vec{x})$-integrations and $(N-1)\ \phi(\vec{x})$-integrations). (See (2.11)) We define (10.65) by N p-integrations and (10.66) by $N-1$ q-integrations matching the time slice with that of the phase space path integral.

Next we make the following change of variables:

$$\phi(x,t) \rightarrow \tilde{\phi}(x,t) = \phi(x + q(t),t) \tag{10.68}$$

$$\pi(x,t) \rightarrow \tilde{\pi}(x,t) = \pi(x + q(t),t) \tag{10.69}$$

We note that the total momentum and the total energy is invariant under this transformation. We also note

$$\frac{\partial Q[\pi(x + q(t),t),\phi(x + q(t),t)]}{\partial q(t)} =$$

$$= \int dx \left(\frac{\partial \pi(x + q(t),t)}{\partial q(t)} \frac{\delta Q[\tilde{\pi},\tilde{\phi};t]}{\delta \tilde{\pi}(x,t)} + \frac{\partial \phi(x + q(t),t)}{\partial q(t)} \frac{\delta Q[\tilde{\pi},\tilde{\phi};t]}{\delta \tilde{\phi}(x,t)} \right.$$

$$= \int dx \left(\tilde{\pi}'(x) \frac{\delta Q}{\delta \tilde{\pi}(x)} + \phi'(x) \frac{\delta Q}{\delta \tilde{\phi}(x)} \right)$$

$$= \int dx \left(\frac{\delta P}{\delta \tilde{\phi}(x)} \frac{\delta Q}{\delta \tilde{\pi}(x)} - \frac{\delta P}{\delta \tilde{\pi}(x)} \frac{\delta Q}{\delta \tilde{\phi}(x)} \right)$$

$$\equiv [P,Q]_p \tag{10.70}$$

where $[.]_p$ is the Poisson bracket with new variables.

Momentum Integration

Next we prove

$$\Psi_I[\phi(\cdot,t_i)] = e^{ip_I q(t_i)} \Psi_I[\tilde{\phi}(\cdot,t_i)] \tag{10.71}$$

where p_I is the momentum of I state.

$$\hat{p} \mid I > = p_I \mid I > \tag{10.72}$$

(proof)

$$\hat{\phi}(x,t) \mid \phi,t > = \mid \phi,t > \phi(x,t) \tag{10.73}$$

$$\hat{\phi}(x,t) \mid \tilde{\phi},t > = \mid \tilde{\phi},t > \tilde{\phi}(x,t) = \mid \tilde{\phi},t > \phi(x + q(t),t) \tag{10.74}$$

Thus,

$$\mid \tilde{\phi},t > = e^{i\hat{p}q(t)} \mid \phi,t > \tag{10.75}$$

$$< \phi,t \mid I > = < \tilde{\phi},t \mid e^{i\hat{p}q(t)} \mid I > = e^{ipq(t)} < \tilde{\phi},t \mid I > \tag{10.76}$$

(QED)

After the change of variables we obtain

$$< F,t_f \mid I,t_i > =$$

$$= \int \cdots \int D\tilde{\phi} D\tilde{\pi} Dp Dq e^{i(p_I q(t_i) - p_F q(t_f))} \Psi_F^*[\tilde{\phi}(\cdot,t_f)] \Psi[\tilde{\phi}\cdot,t_i)] \times$$

$$\times \prod_t \delta(p(t) - P[\tilde{\pi},\tilde{\phi};t])\delta(Q[\tilde{\pi},\tilde{\phi};t])[P,Q]_P \times$$

$$\times \exp i \int_{t_i}^{t_f} dt\, [p(t)\dot{q}(t) + \int dx\, (\tilde{\pi}(x,t)\dot{\tilde{\phi}}(x,t) - H[\tilde{\pi},\tilde{\phi}])] \qquad (10.77)$$

In this path integral $q(t)$ appears only in the exponent as $\int dt p(t)\dot{q}(t)$. So we can integrate it to produce

$$\delta(p_I - p_F)\prod_t \delta(p(t) - p_I) \qquad (10.78)$$

With this δ-function, we can perform the $p(t)$ integration, and we obtain

$$< F, t_f \mid I, t_i > =$$

$$= \delta(p_I - p_F)\int \cdots \int D\tilde{\pi}D\tilde{\phi}\Psi_F^*\Psi_I \prod_t \delta(p - P[\tilde{\pi},\tilde{\phi};t])\, \delta(Q[\tilde{\pi},\tilde{\phi};t])[P,Q]_P \times$$

$$\times \exp i \int\int dx dt\, [\tilde{\pi}\dot{\tilde{\phi}} - H] \qquad (10.79)$$

From now on we use p for $p_I = p_F$.

In order to obtain the stationary paths of this integral we first use

$$\Pi_t \delta(p - P[\tilde{\pi},\tilde{\phi};t]) = \int \cdots \int D\lambda e^{i\int dt\, \lambda(t)(p - P[\tilde{\pi},\tilde{\phi};t])} \qquad (10.80)$$

and then make variations with respect to $\tilde{\pi},\tilde{\phi}$ and λ. We then obtain (10.60) and (10.61). A solution is the following soliton solution:

$$\tilde{\phi}_c(x) = \phi_0\left(\frac{1}{\sqrt{1-v^2}}x\right) \qquad (10.81)$$

$$\tilde{\pi}_c(x) = -v\tilde{\phi}_c{}'(x) = -\frac{v}{\sqrt{1-v^2}}\phi_0{}'\left(\frac{1}{\sqrt{1-v^2}}x\right) \qquad (10.82)$$

In deriving the above solution we did not use the condition $Q=0$. We note $\tilde{\phi}_c(x - x_0)$ is also a solution of (10.60) and (10.61). We choose the condition $Q=0$ in such a way that (10.81) is the only solution, i.e. $x_0=0$. For this purpose we take the form of Q (see (10.67)) as

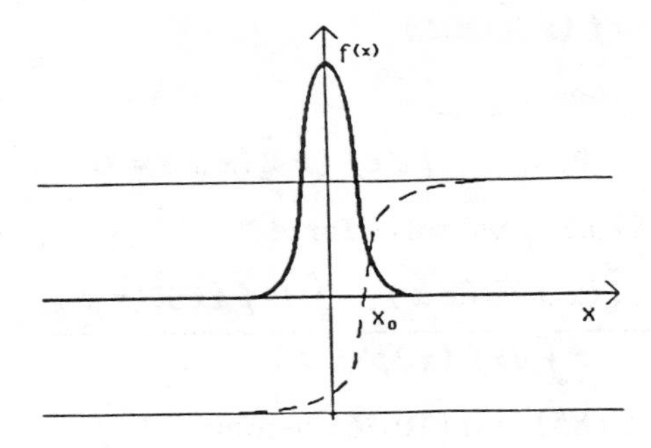

Fig. 10-9

$$Q = \int dx f(x) \tilde{\phi}(x,t)$$

and choose the function $f(x)$ in such a way that when $\tilde{\phi}_c(x - x_0)$ is inserted $Q=0$ is satisfied only when $x_0=0$. Since $\tilde{\phi}_c(x - x_0)$ is a kink solution shown in the dotted line in Fig. 10-9, if we choose $f(x)$ as the zero mode solution of the fluctuation equation

$$f(x) = \frac{E_0^{1/2}}{M_0}\phi_0'\left(\frac{1}{\sqrt{1-v^2}}x\right) \qquad (10.83)$$

which is denoted by the solid line in the same figure, it is quite obvious that $Q = 0$ is satisfied only when $x_0 = 0$.

We normalized f as

$$\int dx f^2(x) = 1 \qquad (10.84)$$

Expansion about Soliton Solution:

Next we expand $\tilde{\phi}$ and $\tilde{\pi}$ around the classified solution. Noticing $\tilde{\pi}_c$ is proportional to $f(x)$, we set

$$\tilde{\phi}(x,t) = \tilde{\phi}_c(x) + \tilde{\eta}(x,t) \qquad (10.85)$$

$$\tilde{\pi}(x,t) = Af(x) + \tilde{\zeta}(x,t) \qquad (10.86)$$

and choose A such that

$$p - P[\tilde{\pi}, \tilde{\phi}] = \int dx f(x) \tilde{\zeta}(x, t) \tag{10.87}$$

Then the constraints become

$$\int dx f(x) \tilde{\eta}(x, t) = 0, \qquad \int dx f(x) \tilde{\zeta}(x, t) = 0 \tag{10.88}$$

Inserting (10.86) into (10.87) we solve for A :

$$A = - \frac{p - \int dx \tilde{\zeta}(x, t) \tilde{\phi}'(x, t) - \int dx f(x) \tilde{\zeta}(x, t)}{\int dx f(x) \tilde{\phi}'(x, t)} \tag{10.89}$$

Thus, the expansions (10.85) and (10.86) become

$$\tilde{\phi}(x, t) = \phi_0(\frac{1}{\sqrt{1 - v^2}} x) + \tilde{\eta}(x, t) \tag{10.90}$$

$$\tilde{\pi}(x, t) = \tilde{\zeta}(x, t) - \phi_0'(\frac{1}{\sqrt{1 - v^2}} x) \frac{p + \int \tilde{\zeta}(x, t) \tilde{\eta}'(x, t) dx}{M_0 + \int \phi_0'(\frac{1}{\sqrt{1 - v^2}} x) \tilde{\eta}'(x, t) dx} \tag{10.91}$$

We then change variables from $\tilde{\pi}, \tilde{\phi}$, to $\tilde{\zeta}, \tilde{\eta}$. This change of variables is a point canonical transformation. As we discussed in **VI** one should add the contribution due to the operator ordering. But, as we mentioned there, it is of order $\hbar^2$. In this lecture we study only the one loop correction (i.e. order of $\hbar$) and disregard the operator ordering.

Let us first compute the Jacobian.

$$J = \det \frac{\delta \tilde{\pi}}{\delta \zeta} = \exp \left(tr \ln \frac{\delta \tilde{\pi}}{\delta \zeta} \right) \tag{10.92}$$

where

$$\frac{\delta \tilde{\pi}(x, t)}{\delta \zeta(x', t)} = \delta(x - x') - \frac{1}{\int f \tilde{\phi}'} f(x) [\tilde{\phi}'(x', t) - f(x')]$$

$$\equiv \langle x | \left(1 - \frac{| f \rangle (\langle \tilde{\phi}' | - \langle f |)}{f \tilde{\phi}'} \right) | x' \rangle \tag{10.93}$$

So,

$$J = \exp tr \left| - \sum_n \frac{1}{n} \left(\frac{|f\ \rangle(\langle\tilde{\Phi}'| - \langle f\ |)}{\int f \tilde{\Phi}'} \right)^n \right|$$

$$= \exp[- \sum_n \frac{1}{n} (\frac{\int f \tilde{\Phi}' - 1}{\int f \tilde{\Phi}'})^n]$$

$$= \exp \ln \left| 1 - \frac{\int f \tilde{\Phi}' - 1}{\int f \tilde{\Phi}'} \right|$$

$$= (\int f \tilde{\Phi}')^{-1} = ([P,Q]_p)^{-1} \qquad (10.94)$$

which cancels with $[P,Q]_P$ in the path integral (10.79).

Next we insert (10.90) and (10.91) into the action and expand it. Since we are interested in the one loop correction we keep the terms only up to quadratic in $\tilde{\zeta}$ and $\tilde{\eta}$. We obtain the following expression for the Hamiltonian

$$\int dxH = E_0 + \frac{1}{2} \int dx [\tilde{\zeta}^2 + \tilde{\eta}'^2 + V''(\phi_0(\frac{1}{\sqrt{1-v^2}} x))\tilde{\eta}^2]$$

$$+ v \int dx \tilde{\zeta} \tilde{\eta} + \frac{3p^2 E_0}{2M_0^4} (\int dx \phi_0''(\frac{1}{\sqrt{1-v^2}} x) \tilde{\eta}(x,t))^2 \qquad (10.95)$$

and

$$\int dx \tilde{\pi}(x,t) \dot{\tilde{\Phi}}(x,t) = \int dx \tilde{\zeta}(x,t) \dot{\tilde{\eta}}(x,t) \qquad (10.96)$$

for the action integral (note (10.88)). Thus, the path integral expression is given by

$$\langle F, t_f \mid I, t_i \rangle =$$

$$= \delta(p_I - p_F) \int \cdots \int D\tilde{\zeta} D\tilde{\eta} \psi_F^* \psi_I \delta(\int f \tilde{\eta}) \delta(\int f \tilde{\zeta}) e^{i \int \int dxdt [\tilde{\zeta} \dot{\tilde{\eta}} - H]}$$

$$(10.97)$$

From the form of action integral (10.96) we see that $\tilde{\zeta}$ and $\tilde{\eta}$ are canonical conjugates of each other. However, in the Hamiltonian (10.95) there is a term which involves a product of $\tilde{\zeta}$ and $\tilde{\eta}$. If we try to get rid of this term by the following transformation:

$$\tilde{\eta}(x\, t) \to \eta(x\, t) = \tilde{\eta}(x - vt\, t)$$

$$\tilde{\zeta}(x\, t) \to \zeta(x\, t) = \tilde{\zeta}(x - vt\, t) \qquad (10.98)$$

then, inside of ϕ_0 in the Hamiltonian becomes time dependent $(\frac{x - vt}{\sqrt{1 - v^2}})$, so that again we can not apply the standard normal mode method. One must invent a generalized time dependent normal mode expansion method such that the action integral is diagonalized at once. This was done for the case of periodic boundary condition by Dashen, Hasslacher and Neveu. (See also Gervais and Jevicki.) Howeer, we do not go into the discussion here because we feel more considerations are needed to apply this method to a single soliton problem.

If this program is completed the corrections to the total energy of the system is obtained as a function of p. One expects it has a form

$$\Delta E_s = \frac{M_0}{E_0} \Delta M \qquad (10.99)$$

so that it is an expansion of the relativistic form

$$E_s = \sqrt{p^2 + (M_0 + \Delta M)^2} \qquad (10.100)$$

We formulated the solitons in quantum field theory by using the Hamiltonian formalism, which is inherently non-covariant. So, it is important to confirm the Lorentz invariance of the theory. This should be done as described above by regarding v as a parameter of order 1. Another way to do this is to use a perturbation series regarding p as a parameter of order 1, so that $v \sim O(g^2)$, then to sum up the series of powers of p. The calculation is tedious but it can be done. In what follows we simply disregard the relativistic problem and discuss only the renormalization of the rest mass of the soliton.

One Loop Quantum Corrections of Soliton Mass and Renormalization

For the case of $v = 0$, i. e. for the soliton at rest, the Hamiltonian (10.95) becomes extremely simple.

$$H = \int dx \frac{1}{2}[\zeta^2 + \eta'^2 + V''(\phi_0(x))\eta^2] \qquad (10.101)$$

Hamilton's equation derived from this is the same as in equation for the small fluctuation (10.34). Thus, by using a set of normalized solutions of (10.34) we can do the normal mode expansion.

$$\zeta(x,t) = \sum_{n \neq 0} p_n(t)\psi_n(x)$$

$$\eta(x,t) = \sum_{n \neq 0} q_n(t)\psi_n(x) \tag{10.102}$$

We obtain

$$H = \sum_{n \neq 0} \frac{1}{2}(p_n^2 + \omega_n^2 q_n^2) \tag{10.103}$$

The absence of the $n = 0$ mode is due to the constraints (10.88)

Now let us go back to the action (10.64). Up until now we neglected the last two terms in the action, namely the renormalization terms δL and the vacuum zero point energy term ΔE. Since the total energy of the system at $v = 0$ is the soliton mass, the mass correction of order g^0 consists of three terms.

The first is of course the main quantum corrections due to (10.103).

$$\Delta M_1 = \frac{1}{2} \sum_{n \neq 0} \omega_n \tag{10.104}$$

The second is due to the renormalization term δL. Here again we shift the field by ϕ_0 and expand it. Since δm^2 is of order g^2 (see (10.48) and since ϕ_0 is of order $1/g$, the order g^2 contribution of δL is given by

$$\Delta M_2 = \frac{1}{4} \delta m^2 \int dx \, (\phi_0^2 - \frac{1}{g^2}) \tag{10.105}$$

The third is the vacuum zero point energy

$$\Delta M_3 = -\frac{1}{2} \sum_n \omega_n^{(0)} \tag{10.106}$$

where $\omega_n^{(0)}$ is the energy of free field, i.e.

$$(-\frac{\partial^2}{\partial x^2} + m^2)\phi_n^{(0)} = \omega_n^{(0)2} \phi_n^{(0)} \tag{10.107}$$

The total corrections are the sum of these terms. Since each of them contains an infinity, the sum should be taken carefully. We note that the potential $U(x)$ in (10.34) approaches m^2 for large x so that the density of the states of the fluctuation modes in high momentum is the same as that of the free case. So, in the sum of $\Delta M_1 + \Delta M_3$ the main (linear) divergences would cancel each other. In order to see the cancellations in what follows we regularize the expression by putting the system in a box of length L.

For the potential of (10.41), the equation for the small fluctuation (10.34) is given by

$$\left[-\frac{\partial^2}{\partial x^2} - \frac{m^2}{2} + \frac{3m^2 g^2}{2}\phi_0^2(x)\right]\psi(x) = \omega^2\psi(x) \tag{10.108}$$

where

$$\phi_0(x) = \frac{1}{g}\tanh\frac{mx}{2} \tag{10.109}$$

There are two bound state solutions. One is the zero energy solution relevant to the center of mass motion of the kink which we discussed earlier.

$$\psi_0(x) = \frac{1}{\sqrt{M_0}}\phi_0'(x) \tag{10.110}$$

where M_0 is the classical mass of the soliton given by

$$M_0 = \frac{m}{6g^2} \tag{10.111}$$

The other is given by

$$\psi_1(x) = \frac{1}{N_1}\frac{\sinh\frac{mx}{2}}{\cosh^2\frac{mx}{2}}, \qquad \omega_1 = \frac{\sqrt{3}m}{2} \tag{10.112}$$

which corresponds to some eigenvibration of the kink.

Since the potential in (10.108) approaches to m^2 exponentially as x approaches infinity, we have continuum spectra above m. The corresponding energy and wave function are given by

$$\omega(k) = (k^2 + m^2)^{1/2}$$

$$\psi_q(x) = \frac{1}{N_q} e^{iqx} [3\tanh\frac{mx}{2} - 6i\frac{q}{m}\tanh\frac{mx}{2} - 1 - 4\frac{s^2}{m^2}] \quad (10.113)$$

This is a traveling wave solution with no reflections. We put this solution in the box with periodic boundary conditions. Then momenta are quantized by

$$q_n L + \delta(q_n) = 2\pi n \quad (10.114)$$

where the phase shift $\delta(q)$ is defined by

$$\psi_q(\pm L/2) \underset{L\to\infty}{\longrightarrow} \exp(\pm i[qL/2 + \frac{1}{2}\delta(q)]) \quad (10.115)$$

and found to be

$$\delta(q) = \begin{cases} 2\pi - 2\tan^{-1}(2q/m) - 2\tan^{-1}(q/m) & q > 0 \\ -2\pi - 2\tan^{-1}(2q/m) - 2\tan^{-1}(q/m) & q < 0 \end{cases} \quad (10.116)$$

On the other hand the momenta of free field are quantized as

$$k_n = 2\pi n / L, \qquad \omega_n^{(0)2} = k_n^2 + m^2 \quad (10.117)$$

Since there are two bound state solutions for the small fluctuation equation (10.108). we regard them as the two lowest free field states moved to the bound states with the presence of the kink.. So the sum of ΔM_1 and ΔM_3 is given by

$$\sum_n \frac{1}{2}\omega_n - \sum_n \frac{1}{2}\omega(k_n) = \frac{1}{2}(\omega_0 - m) + \frac{1}{2}(\omega_1 - m) + \frac{1}{2}\sum_n(\omega(q_n) - \omega(k_n))$$

$$(10.118)$$

which is approximated by

$$\frac{1}{2}\omega_1 - m + \frac{1}{2}[\omega(k_n - \frac{1}{L}\delta(k_n)) - \omega(k_n)] \quad (10.119)$$

$$\approx \frac{1}{2}\omega_1 - m - \frac{1}{2L}\sum_n \frac{d\,\omega(k_n)}{dk_n}\delta(k_n) \quad (10.120)$$

The last term is written in an integral form as

$$-\frac{1}{4\pi}\int dk\,\delta(k)\frac{d\,\omega(k)}{dk} = -\frac{1}{4\pi}[\omega(k)\delta(k)]_{-\infty}^{\infty} + \frac{1}{4\pi}\int dk\frac{d\,\delta(k)}{dk}\omega(k) =$$

$$= -\frac{3m}{2\pi} - \frac{3m}{2\pi}\int\frac{m^2\,dk}{(k^2+m^2/4)(k^2+m^2)^{1/2}} - \frac{3m}{2\pi}\int_0^{\Lambda}\frac{dk}{(k^2+m^2)^{1/2}}$$

(10.121)

Obviously the last term diverges logarithmically, but note the leading linear divergences were canceled.

Now let us look at ΔM_2 (10.105). Inserting the expression for δm^2 (10.48) and ϕ_0 (10.109) into (10.105) and regularizing the divergent integral Δ with the box again we find that ΔM_2 is exactly the same as the divergent term in (10.110) with the opposite sign. And we obtain

$$\Delta M = m\left(\frac{1}{4\sqrt{3}} - \frac{3}{2\pi}\right) \tag{10.122}$$

Exercise: Complete the calculation.

AI. Quantum Theory of Non-Abelian Gauge Fields

A1.1 Classical Gauge Field Theory

QED

QED Lagrangian:

$$L = -\frac{1}{4}F_{\mu\nu}F^{\mu\nu} + \bar{\psi}(i\gamma\cdot\partial - m)\psi - ie\bar{\psi}\gamma^{\mu}\psi A_{\mu} \qquad \text{(A1.1)}$$

$$F_{\mu\nu} = \partial_{\mu}A_{\nu} - \partial_{\nu}A_{\mu} \qquad \text{(A1.2)}$$

Using the covariant derivative notation defined by

$$D_{\mu} = \partial_{\mu} - ieA_{\mu} \qquad \text{(A1.3)}$$

one can combine the second and third term of L as

$$\bar{\psi}(i\gamma\cdot D - m)\psi \qquad \text{(A1.4)}$$

Note

$$[D_{\mu}, D_{\nu}] \equiv D_{\mu}D_{\nu} - D_{\nu}D_{\mu} = -ieF_{\mu\nu} \qquad \text{(A1.5)}$$

Gauge transformations:

$$\psi(x) \rightarrow e^{ie\Lambda}\psi(x) \qquad (first\ kind) \qquad \text{(A1.6)}$$

$$A_{\mu}(x) \rightarrow A_{\mu}(x) + \partial_{\mu}\Lambda(x) \qquad (second\ kind) \qquad \text{(A1.7)}$$

Under the gauge transformations, $F_{\mu\nu}$ is invariant and $D_{\mu}\psi$ transforms as ψ.

At first sight the 2nd kind of gauge transformation does not have an intimate relation with the lst kind, but if one defines the following path dependent phase factor

$$u(x, x') = \exp ie\int_{x}^{x'} dx^{\mu}A_{\mu}(x) \qquad \text{(A1.8)}$$

it transforms under gauge transformation as

$$u(x,x') \rightarrow e^{ie\Lambda(x)}u(x,x')e^{-ie\Lambda(x')} \tag{A1.9}$$

Namely $u(x,x')$ transforms as $\psi(x)\bar{\psi}(x')$. Note

$$u(x,x')\Big|_{x=x'} = 1 \tag{A1.10}$$

so that

$$A_\mu(x) = \frac{1}{ie}\partial_\mu u(x,x')\Big|_{x=x'} \tag{A1.11}$$

The covariant derivative transforms as

$$D_\mu(x) \rightarrow e^{ie\Lambda(x)}D_\mu(x)e^{-ie\Lambda(x)} \tag{A1.12}$$

so that

$$F_{\mu\nu} = \frac{i}{e}[D_\mu,D_\nu] \rightarrow \frac{i}{e}e^{ie\Lambda(x)}[D_\mu,D_\nu]e^{-ie\Lambda(x)}$$

$$= e^{ie\Lambda(x)}F_{\mu\nu}e^{-ie\Lambda(x)} \tag{A1.13}$$

Yang-Mills Field Theory

Since $e^{ie\Lambda(x)}$ is an element of U(1) we call the gauge theory of the previous section (QED) U(1) gauge theory. In order to extend U(1) gauge theory to SU(N), one considers a $\psi(x)$ which is a fundamental representation of SU(N):

$\psi_{i\alpha}$: $\alpha = 1, 2, \cdots, N$ (*color index*) $\quad i = 1, 2, 3, 4$ (*spinor index*)

Let t^a be a fundamental representation (NxN) of SU(N) generators:

$$[t^a, t^b] = if_{abc}t^c \tag{A1.14}$$

where the index runs $1,2,\cdots,N^2-1$. One adds $t^0 = \frac{1}{\sqrt{N}}\mathbf{1}$ to this set and calls it a U(N) algebra. The fundamental representation of a U(N) algebra has the following important properties:

$$tr(t^a t^b) = \delta_{ab} \tag{A1.15}$$

$$\sum_{a=0}^{N^2-1}(t^a)_{\alpha\beta}(t^a)_{\alpha'\beta'} = \delta_{\alpha\beta'}\delta_{\alpha'\beta} \tag{A1.16}$$

Caution: Our normalization of t^a is different from the usual one, it is larger by factor $\sqrt{2}$. Accordingly, the structure constant f_{abc} in (A1.14) is also larger by a factor of $\sqrt{2}$.

The elements of U(N) are parametrized as

$$U(\Omega) = \exp igt^a \Omega_a \tag{A1.17}$$

where Ω_a $(a = 0,1, \cdots N^2 - 1)$ are real parameters.

Gauge transformations:

$$\psi(x) \rightarrow \psi^{\Omega}(x) = U(\Omega(x))\psi(x) \tag{A1.18}$$

$$u(x,x') \rightarrow u^{\Omega}(x,x') = U(\Omega(x))u(x,x')U^{+}(\Omega(x')) \tag{A1.19}$$

The vector potential is defined by

$$A_{\mu}(x) = \frac{1}{ig}\frac{\partial}{\partial x_{\mu}}u(x,x')\Big|_{x=x'} \tag{A1.20}$$

so

$$A_{\mu} \rightarrow A_{\mu}^{\Omega}(x) = U(\Omega(x))A_{\mu}(x)U^{+}(\Omega(x)) - \frac{i}{g}(\partial_{\mu}U(\Omega(x)))U^{+}(\Omega(x)) \tag{A1.21}$$

The covariant derivative is defined in such a way that under gauge transformations it transforms as

$$D_{\mu}(A^{\Omega}) = U(\Omega(x))D_{\mu}(x)U^{+}(\Omega(x)) \tag{A1.22}$$

The following definition is compatible with this requirement.

$$D_{\mu}(A) = \partial_{\mu} - igA_{\mu}(x) \tag{A1.23}$$

$F_{\mu\nu}$ is defined by (see (A1.15))

$$F_{\mu\nu} = \frac{i}{g}[D_{\mu},D_{\nu}] = \partial_{\mu}A_{\nu} - \partial_{\nu}A_{\mu} - ig[A_{\mu},A_{\nu}] \tag{A1.24}$$

accordingly,

$$F_{\mu\nu} \rightarrow F_{\mu\nu}^{\Omega} = U(\Omega)F_{\mu\nu}U^{+}(\Omega) \tag{A1.25}$$

A gauge invariant Lagrangian, which is the simplest generalization of

(A1.1) is given by

$$L = -\frac{1}{4} tr\,(F_{\mu\nu}F^{\mu\nu}) + \bar{\psi}(i\gamma\cdot D + m)\psi \tag{A1.26}$$

A1.2 Quantum Theory of Yang-Mills Field

$A_0 = 0$ **gauge:**

$$L = -\frac{1}{4}F^a_{\mu\nu}F^{a\mu\nu} \tag{A1.27}$$

$$F^a_{\mu\nu} = \partial_\mu A^a_\nu - \partial_\nu A^a_\mu + g f_{abc} A^b_\mu A^c_\nu \tag{A1.28}$$

$$A_\mu \equiv A^a_\mu t^a \tag{A1.29}$$

For a given A_μ , there exists a time dependent gauge transformation Ω such that $A_0^\Omega = 0$.

(proof)

$$U(\Omega)A_0U^+(\Omega) + \frac{i}{g}U(\Omega)\dot{U}^+(\Omega) = 0$$

The explicit solution is

$$U^+(\Omega) = T\exp ig\int_0^t dt'A_0(\vec{x},t')$$

(QED)

In this gauge the Lagrangian density is given by

$$L = \frac{1}{2}(\dot{A}^a_i\dot{A}^a_i - B^a_iB^a_i) \tag{A1.30}$$

where

$$B^a_i = \frac{1}{2}\epsilon_{ijk}F^a_{jk} \tag{A1.31}$$

Canonical Formalism:

Compare (A1.30) with the standard form

$$L = \frac{1}{2}\sum_i \dot{q}_i^2 - V(q)$$

$$q^i \longleftrightarrow A_i^a(\vec{x}) \qquad i \longleftrightarrow i, a, \vec{x} \tag{A1.32}$$

Symmetry:

Potential model: O (N) rotational symmetry

Yang-Mills: time independent non-Abelian gauge symmetry

Non-relativistic potential model:

Let us consider infinitesimal O(N) transformation

$$\delta x_i = \epsilon_{ij} x_j \qquad \epsilon_{ij} = -\epsilon_{ji} \tag{A1.33}$$

We assume the potential is invariant (central).

$$V(x + \delta x) = V(x) \tag{A1.34}$$

$$\delta L = \sum_i \dot{x}_i \delta \dot{x}_i = \dot{\epsilon}_{ij} \dot{x}_i x_j$$

So, if

$$\dot{\epsilon}_{ij} = 0,$$

then

$$\delta L = 0$$

$$\delta \int dt L = -\int dt\, \dot{\epsilon}_{ij} L_{ij} = \int dt\, \epsilon_{ij} \dot{L}_{ij}$$

where

$$L_{ij} = \frac{1}{2}(x_i \dot{x}_j - \dot{x}_i x_j) \tag{A1.35}$$

Noether's theorem:

$$\dot{L}_{ij} = 0 \tag{A1.36}$$

Canonical formalism:

$$p_i = \partial L / \partial \dot{x}_i = \dot{x}_i \tag{A1.37}$$

$$H = \sum_i \dot{x}_i p_i - L = \frac{1}{2}\sum_i p_i^2 + V(x) \tag{A1.38}$$

$$L_{ij} = \frac{1}{2}(x_i p_j - x_j p_i) \tag{A1.39}$$

$$[H, L_{ij}]_P = 0. \qquad [\ ,\ ]_P \ : \textit{Poisson bracket} \tag{A1.40}$$

$$\delta x_i = [\frac{1}{2}\epsilon_{lj} L_{lj}, x_i]_P = \epsilon_{ij} x_j \tag{A1.41}$$

generator of infinitesimal transformation.

Yang-Mills Theory:

Canonical momentum:

$$E_i^a(\vec{x}) = \frac{\partial L}{\partial \dot{A}_i^a(\vec{x})} = \dot{A}_i^a(\vec{x}) \tag{A1.42}$$

Hamiltonian

$$H = \int d\vec{x}[\frac{1}{2}(E_i^a(\vec{x}))^2 + \frac{1}{2}(B_i^a(\vec{x}))^2] \tag{A1.43}$$

potential term

The potential term is gauge invariant.

Infinitesimal gauge transformation:

$$\delta A_i^a(\vec{x}) = \partial_i \Omega^a(\vec{x}) + g f_{abc} A_i^b(\vec{x})\Omega^c(\vec{x}) \tag{A1.44}$$

$$\delta \int L dt = - \int dt \int d\vec{x}\, \Omega^a(\vec{x}) P_a(\vec{x}) \tag{A1.45}$$

where

$$P_a(\vec{x}) = \partial_i E_i^a(\vec{x}) + g f_{abc} A_i^b(\vec{x}) E_i^c(\vec{x}) \tag{A1.46}$$

Classical Gauss' theorem:

$$P_a(\vec{x}) = 0 \qquad \textbf{(A1.47)}$$

Quantization:

In the following we define the quantum theory of non-Abelian gauge fields in the $A_0 = 0$ canonical formalism.

$$E_i^a(\vec{x}) = -i\hbar \frac{\partial}{\partial A_i^a(\vec{x})} \qquad (A1.48)$$

and the subsidiary condition:

$$P_a(\vec{x}) \mid \Psi > = 0 \qquad (A1.49)$$

II.3 Equivalence of $A_0 = 0$ canonical quantization and covariant quantization

The condition (A1.49) is equivalent to

$$\Psi[A^{\Omega}] = \Psi[A] \qquad (A1.50)$$

where

$$\Psi[A] = <A \mid \Psi> \qquad (A1.51)$$

The transition amplitude defined by

$$T_{fi} = <\Psi_f(t_f) \mid \Psi_i(t_i)> = <\Psi_f \mid e^{-iH(t_f - t_i)} \mid \Psi_i>$$

is expressed in a form of path integral by using the technique of II and III As before

$$t_f - t_i = N\epsilon$$

$$t_n = t_i + n\epsilon$$

then insert

$$\int dA \mid A><A \mid = 1$$

successively, we obtain

$$T_{fi} = \int \cdots \int \prod_{n=0}^{N} dA(n) \Psi^*{}_f[A(N)]\, \Psi_i[A(0)] \prod_{n=0}^{N-1} <A(n+1) \mid e^{-iH\epsilon} \mid A(n)>$$

We use the following abbreviation throughout this section.

$$A(n) = A_i^a(\vec{x}, t_n)$$

By using

$$\langle A(n+1) | e^{-i\epsilon H} | A(n) \rangle =$$

$$= \int dE(n)(1 - i\frac{\epsilon}{2}\int dx\,(E(n)^2 + B(n)^2))e^{iE(n)(A(n+1)-A(n))}$$

$$= const. \int dE(n) \exp i\epsilon \int dx\,[\epsilon^{-1}E(n)(A(n+1)-A(n)) - \tfrac{1}{2}(E(n)^2 + B(n)^2)]$$

$$= const. \exp i\frac{1}{2}\int dx\,[(A(n+1)-A(n))^2\epsilon^{-1} - \epsilon B^2(n)]$$

Accordingly,

$$T_{fi} = \int \cdots \int \prod_{n=0}^{N} dA(n)\, \Psi^*_f[A(N)]\Psi_i[A(0)] \times$$

$$\times \exp i\frac{1}{2}\int d\vec{x} \sum_{n=0}^{N-1} [(A(n+1)-A(n))^2\epsilon^{-1} - \epsilon B^2(n)] \qquad \text{(A1.52)}$$

where we simply wrote $B^2(n)$ for $trB^2(\vec{x}, t_n)$.

Under gauge transformation

$$(B^{\Omega}(n))^2 = B^2(n)$$

and also the measure of integration is invariant.

(proof)

Using (A1.21) we obtain

$$d(A^{\Omega})^a = tr(t^a U(\Omega) t^b U^+(\Omega)) dA^b = \Lambda_{ab} dA^b$$

$$\Lambda_{ab}\Lambda_{cb} = \delta_{ac}$$

since

$$tr(t^a U t^b U^+) tr(t^c U t^b U^+) = tr(U^+ t^a U U^+ t^c U) = \delta_{ac}$$

thus,

$$(\det\Lambda)^2 = 1 \rightarrow \det\Lambda = 1$$

by continuity (note for $U=1$ $\Lambda=1$). Thus,

$$dA^{\Omega} = dA$$

(QED)

Faddeev-Popov trick:

$$\int \prod_{n=1}^{N-1} dA_0(n)\delta(\nabla A^{\Omega^{-1}}(n) + \frac{1}{\epsilon}(A_0(n+1) - A_0(n)))\, J = 1 \qquad \text{(A1.53)}$$

where we relate A_0 and Ω by

$$e^{ig\epsilon A_0(n)} = U^+(\Omega(n))U(\Omega(n+1)) \qquad \text{(A1.54)}$$

Inserting these relations into the path integral expression (A1.52) and changing the integration variables from A to A^{Ω} we obtain

$$T_{fi} = \int \cdots \int \prod_{n=0}^{N} dA(n) \prod_{n=0}^{N-1} dA_0(n) \Psi^*_f[A(N)]\Psi_i[A(0)]\, J \times$$

$$\times \prod_{n=0}^{N-1} \delta(\nabla A(n) + \frac{1}{\epsilon}(A_0(n+1) - A_0(n)))$$

$$\times \exp i\frac{1}{2}\sum_{n=0}^{N-1} \int d\vec{x}[(A^{\Omega}(n+1) - A^{\Omega}(n))^2\frac{1}{\epsilon} - \epsilon B^2(n)] \qquad \text{(A1.55)}$$

And

$$\frac{1}{\epsilon}tr(A^{\Omega}(n+1) - A^{\Omega}(n))^2$$

$$= \epsilon tr[(A(n+1) - A(n))\frac{1}{\epsilon} + ig[A_0(n),A(n)] + \nabla A_0(n)]^2 + O(\epsilon^2)$$

$$\xrightarrow[\epsilon \to 0]{} \epsilon tr F_{0i}^2$$

Thus,

$$T_{fi} = \int \cdots \int DA_\mu \Psi^*_f[A(\cdot,t_f)]\Psi_i[A(\cdot,t_i)] \times$$

$$\times J\ \delta(\partial_\mu A_\mu)\exp i\int d^4x(-\frac{1}{4}trF_{\mu\nu}^2) \qquad \text{(A1.56)}$$

if one can prove

$$J = \det(\partial_\mu D_\mu) \tag{A1.57}$$

the expression (A1.56) is the expression used for the covariant perturbation theory in Landau gauge.

AII. Spin System and Lattice Gauge Theory

A2.1 O(N) Heisenberg Spin System.

Action:

$$S = \frac{1}{2g}\int dtdx\,(\partial_\mu \vec{n}(x,t))^2 \quad (A2.1)$$

$$(\vec{n}(x,t))^2 = 1 \quad (A2.2)$$

O(2):

$$n_1^2(x) + n_2^2(x) = 1 \quad (A2.3)$$

$$n_1(x) = \cos\theta(x), \qquad n_2(x) = \sin\theta(x) \quad (A2.4)$$

θ satisfies

$$\pi \geqslant \theta(x) \geqslant -\pi \quad (A2.5)$$

Thus,

$$L = \frac{1}{2g}[\dot{\theta}^2 - \theta'^2] = \frac{1}{2g}[\dot{\theta}^2 - (\partial_x \vec{n})^2] \quad (A2.6)$$

Canonical momentum:

$$p_\theta = \frac{\partial L}{\partial \dot{\theta}} = \frac{1}{g}\dot{\theta} \quad (A2.7)$$

Hamiltonian:

$$H = \int dx \frac{1}{2g}[\dot{\theta}^2 + (\partial_x \vec{n})^2] \quad (A2.8)$$

To latticize we use the following correspondence:

$$\partial_x \vec{n} \rightarrow \frac{1}{a}[\vec{n}(n+1) - \vec{n}(n)] \quad (A2.9)$$

$$\int dx \to \sum_n a \tag{A2.10}$$

where a is the lattice distance.

$$H = \frac{a}{2g}\sum_m[\dot{\theta}(m)^2 - \frac{2}{a^2}\vec{n}(m)\cdot\vec{n}(m+1)] + const. \tag{A2.11}$$

Angular momentum of O(2):

$$J(m) = \frac{a}{g}\dot{\theta}(m) \tag{A2.12}$$

Thus,

$$H = \frac{g}{2a}\sum_m[J^2(m) - \frac{2}{g^2}\vec{n}(m)\cdot\vec{n}(m+1)] \tag{A2.13}$$

Local O(N) generator: $J_{ij}(m) = -J_{ji}(m)$

$$[J_{ij}(m), n_k(m')] = i(\delta_{ik}n_j(m) - \delta_{jk}n_i(m))\delta_{mm'} \tag{A2.14}$$

This commutation relation and the commutation relations among J's are formally obtained by using

$$J_{ij}(m) = -i[n_i(m)\frac{\partial}{\partial n_j(m)} - n_j(m)\frac{\partial}{\partial n_i(m)}] \tag{A2.15}$$

although rigorously speaking one can not write J in this form since $\vec{n}^2 = 1$.

$$H = \frac{g}{2a}\sum_m[\frac{1}{2}J_{ij}^2(m) - \frac{2}{g^2}\vec{n}(m)\cdot\vec{n}(m+1)] \tag{A2.16}$$

Total angular momentum:

$$J_{ij} = \sum_m J_{ij}(m) \tag{A2.17}$$

A2.2 SU(N) Symmetric Hermitian Matrix Model

N x N Hermitian matrix:

$$M^+ = M \tag{A2.18}$$

Lagrangian:

$$L = \frac{1}{2} tr \dot{M}^2 - V(M) \tag{A2.19}$$

$$V(M) = \frac{1}{2} tr M^2 + \frac{g}{N} tr M^4 \tag{A2.20}$$

Symmetry transformation:

$$M \rightarrow UMU^+ \tag{A2.21}$$

i.e. SU(N) symmetry.

$$M = M_a t^a \tag{A2.22}$$

M_a is a real variable. Since

$$tr \dot{M}^2 = \sum_{a=0}^{N^2-1} \dot{M}_a^2 \tag{A2.23}$$

Lagrangian (A2.19) is of standard form.

$$H = -\frac{1}{2} \sum_a \frac{\partial^2}{\partial M_a^2} + V(M) \tag{A2.24}$$

A2.3 SU(N) Matrix Model (Chiral Model)

Dynamical variable : N x N unitary matrix u

$$uu^+ = 1, \quad u^+ u = 1 \tag{A2.25}$$

Lagrangian:

$$L = \frac{1}{2} tr (\dot{u}\dot{u}^+) \tag{A2.26}$$

Symmetry transformation:

$$u \rightarrow U_L \; u \; U_R \tag{A2.27}$$

Parametrization:

$$u = \exp i t^a A_a \tag{A2.28}$$

$$\frac{\partial u}{\partial A_a} = in_{ab}(A)t^b u \qquad (definition\ of\ n_{ab}) \tag{A2.29}$$

$$n_{ab}(A) = \int_0^1 d\alpha tr(e^{i\alpha A}t^a e^{-i\alpha A}t^b)$$

$$\dot{u} = i\dot{A}_a n_{ab}(A)t^b u \tag{A2.30}$$

$$L = \frac{1}{2}tr(\dot{u}\dot{u}^+) = \frac{1}{2}\dot{A}_a \dot{A}_a n_{ab}(A)n_{a'b}(A) \tag{A2.31}$$

Canonical momentum:

$$\pi_a = \frac{\partial L}{\partial A_a} = \dot{A}_c n_{ab}(A)n_{cb}(A) \tag{A2.32}$$

Define

$$E_a = n_{ab}^{-1}\pi_b = \dot{A}_c n_{ca}(A) \tag{A2.33}$$

Quantum Theory:

$$E_a = n_{ab}^{-1}(-i\frac{\partial}{\partial A_b}) \tag{A2.34}$$

So,

$$[E_a, E_b] = if_{abc}E_c \tag{A2.35}$$

$$[E_a, u] = t_a u \tag{A2.36}$$

Hamiltonian:

$$H = \frac{1}{2}\sum_{a=0}^{N^2-1} E_a^2 \tag{A2.37}$$

Proof of (A2.35):

$$[E_a, E_b] = [n_{ac}^{-1}(-i\frac{\partial}{\partial A_c}), n_{bd}(-i\frac{\partial}{\partial A_d})]$$

$$= in_{ac}^{-1}n_{bd}^{-1}[\frac{\partial}{\partial A_c}n_{de} - \frac{\partial}{\partial A_d}n_{ce}]E_e \tag{A2.38}$$

thus, if

$$\frac{\partial}{\partial A_a} n_{bc} - \frac{\partial}{\partial A_b} n_{ac} = n_{ad} n_{be} f_{dec} \qquad \text{(A2.39)}$$

is satisfied, we obtain (A2.35). To prove it we take a derivative of (A2.28)

$$\frac{\partial}{\partial A_b} \frac{\partial}{\partial A_a} u(A) = i\left(\frac{\partial}{\partial A_b} n_{ab}(A)\right) t^b u(A) - n_{ac} n_{bd} t^c t^d u(A)$$

then interchange a and b and take a difference

$$i\left(\frac{\partial}{\partial A_b} n_{ac} - \frac{\partial}{\partial A_a} n_{bc}\right) t^c u = n_{ad} n_{be} [t^d, t^e] u$$

$$if_{dec} t^c$$

(QED)

A2.4 SU(N) Gauge Theory: Kogut-Susskind Model

Dynamical variable:

$$P \exp ig \int_x^{x'} A_i \, dx^i \sim e^{igaA_i}$$

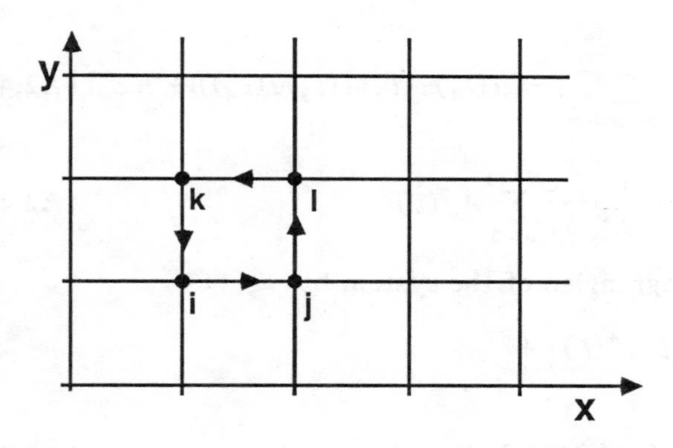

On each link of the square lattice one considers an N x N unitary matrix. They are the dynamical variables of the theory.

$$u(i,j) = e^{igaA_x(i)} \tag{A2.40}$$

$$u(k,i) = u^{+}(i,k) = e^{-igaA_y(i)} \tag{A2.41}$$

etc.

$$tr(u(i,j)\,u(j,l)\,u(l,k)\,u(k,i)) \equiv \square$$

$$= tr\left[e^{igaA_x(i)}\, e^{igaA_y(j)}\, e^{-igaA_x(k)}\, e^{-igaA_y(i)}\right]$$

Using

$$e^A e^B = e^{A + B + ½[A,B] + O(A^3, AB^2, \ldots)}$$

$$\square = tr\left(e^{iga^2\left[\frac{A_y(j)-A_y(i)}{a} - \frac{A_x(k)-A_x(i)}{a} + ig[A_x, A_y]\right]}\right)$$

$$F_{xy}$$

$$= tre^{-iga^2 F_{xy}} = N - \frac{1}{2} g^2 a^4 tr(F_{xy})^2$$

Thus,

$$\frac{1}{4}\int d\vec{x} tr(F_{ij})^2) =$$

$$= -\frac{1}{2g^2 a} \sum_{all\ plaquettes} [tr(u(l_1)u(l_2)u(l_3)u(l_4)) + h.c.] \tag{A2.42}$$

$$\frac{1}{2} tr\left(\dot{u}(l)\dot{u}^{+}(l)\right) = \frac{1}{2} g^2 a^2 \sum_{a=0}^{N^2-1} \dot{A}_a^2(l) \tag{A2.43}$$

Thus, we define the Lagrangian of the system by

$$L = \frac{a}{2g^2} \sum_{l} tr\left(\dot{u}(l)\dot{u}^{+}(l)\right) +$$

$$+ \frac{1}{2g^2 a} \sum_{\gamma} [tr(u(l_1)u(l_2)u(l_3)u(l_4)) + h.c.\ ] \tag{A2.44}$$

Note that the kinetic energy term is the same as SU(N) matrix model (chiral model). So one obtains

$$H = \frac{g^2}{2a}[\sum_l E_a^2(l) - \frac{1}{g^4}\sum_\gamma (tr\,(u(l_1)u(l_2)u(l_3)u(l_4)) + h.c.)] \quad (A2.45)$$

Corresponding to (A2.36) one has

$$[E_a(l), u(l')] = \delta_{ll'} t^a u(l) \quad (A2.46)$$

and to (A2.26)

$$u(l) = u(i,j) \longrightarrow U(i)u(i,j)U^+(j) \quad (A2.47)$$

which we call a gauge transformation. The gauge transformations are the transformations defined on the lattice sites. The K-S Hamiltonian is invariant under gauge transformations.

A2.5 Strong Coupling Expansion

The characteristics of the lattice Hamiltonians are:. i) the kinetic energy term is given by (generalized) angular momentum square . ii) potential term is $O(1/(coupling\ const.)^2)$ compared to the kinetic energy term:

$$H = \frac{g}{2a}W, \qquad W = W_0 - \alpha V, \qquad \alpha = \frac{2}{g^2} \quad (A2.48)$$

where

$$W_0 = \sum_m J^2(m) \quad (A2.49)$$

and

$$V = \frac{1}{2}\sum_m [\phi^+(m)\phi(m+1) + h.c.\ ] \quad (A2.50)$$

$$\phi(m) = n_1(m) + n_2(m) = e^{i\theta} \quad (A2.51)$$

Commutation relations:

$$[J(m), \phi(m')] = \phi(m)\delta_{mm'} \quad (A2.52)$$

$$[J(m), \phi^+(m')] = -\phi^+(m)\delta_{mm'} \quad (A2.53)$$

Diagonalization of W_0:

$$J^2(m)|0\rangle = 0 \tag{A2.54}$$

1st excited state (translationally invariant zero momentum state)

$$|1\rangle = \frac{1}{M}\sum_m \phi^+(m)\ |0\rangle \tag{A2.55}$$

where M = # of lattice sites

$$W_0|1\rangle = |1\rangle \tag{A2.56}$$

Perturbation expansion (strong coupling expansion):

$$W\|0\rangle = \omega_0\|0\rangle, \quad W\|1\rangle = \omega_1\|1\rangle \tag{A2.57}$$

The strong coupling perturbation expansion is obtained as follows.

$$\omega_0 = 0 + \alpha^2 \langle 0|V(0-H)^{-1}V|0\rangle + \ldots$$

$$\sim (\alpha/2)^2 \frac{2M}{0-2} + \ldots\ldots \tag{A2.58}$$

$$\omega_1 = 1 - \alpha\langle 1|V|1\rangle + \cdots$$

$$\sim 1 - \alpha + \cdots \tag{A2.59}$$

Thus,

$$\omega_1 - \omega_0 = F(\alpha)$$

$$\sim 1 - \alpha + O(\alpha^2) \tag{A2.60}$$

A2.6 Renormalization and the β function

Mass gap:

$$E = \frac{g}{2a}(\omega_1 - \omega_0) = \frac{g}{2a}F(\alpha) \tag{A2.61}$$

Since the mass gap is a physical quantity, it should be independent of the choice of lattice distance. The apparent dependence should be canceled by the dependence of coupling constant.

$$\frac{dE}{da} = 0 \tag{A2.62}$$

β function:

$$\frac{\beta(g)}{g} = a\frac{dg}{da} \tag{A2.63}$$

$$\frac{\beta(g)}{g} = \frac{F(\alpha)}{F(\alpha) - 2\alpha F'(\alpha)} \tag{A2.64}$$

Use (A2.60)

$$\frac{\beta(g)}{g} \sim \frac{1-\alpha}{1+\alpha} \sim 1 \qquad \text{(A2.65))}$$

for strong coupling. For weak coupling

$$\beta(g) = 0$$

because, by the scale transformation $\tilde{\theta} = \theta / g$ we can remove g from the Lagrangian and the range of $\tilde{\theta}$ is given by

$$\frac{\pi}{g} \geqslant \tilde{\theta} \geqslant -\frac{\pi}{g}$$

so that in the $g \to 0$ limit the theory becomes a free field theory.

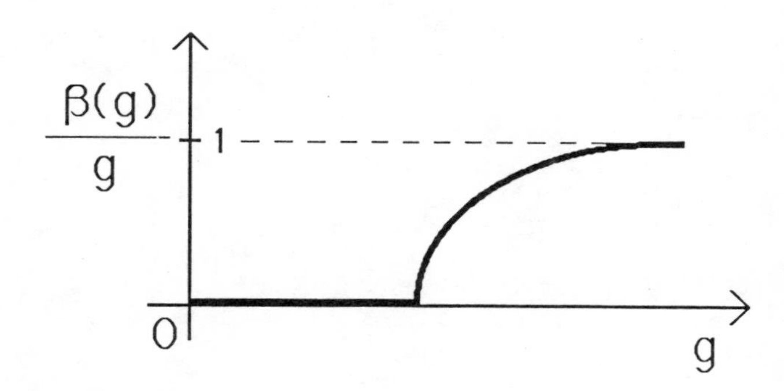

STOCHASTIC QUANTIZATION*

*Supported in part by a grant from the National Science Foundation under contract NSF-PHY-82-15364 and by CCNY Faculty Research Award Program #6-63264.

I. INTRODUCTION

The quantization methods used in field theories are i) canonical, ii) path-integral, and iii) stochastic quantization. The last method is relatively new. It was originated by Parisi and Wu[1)] in 1980. In this lecture I will try to review the stochastic quantization method without assuming reader's knowledge on this subject at all. Thus, I first recall the standard canonical quantization.

Canonical Quantization:

Let $L(q,\dot{q})$ be a Lagrangian of a system, q being a dynamical variable and $\dot{q}$ its time derivative. The canonical momentum p is defined by

$$p = \frac{\partial L}{\partial \dot{q}} \tag{1.1}$$

and Hamiltonian of the system is given by the following Lagrange transform:

$$H(p,q) = p\dot{q} - L(q,\dot{q}) \tag{1.2}$$

In canonical operator formalism of quantum mechanics the dynamical variable $\hat{q}$ and its canonical conjugate momentum $\hat{p}$ are operators in a Hilbert space (from here on operators are denoted with ^) and satisfy the following canonical commutation relation:

$$[\hat{q}, \hat{p}] = i \tag{1.3}$$

The state of the system is a time dependent vector in the Hilbert space (Schrödinger picture), and the mechanical equation of the state vector is the Schrödinger equation:

$$i \frac{\partial}{\partial t} |\psi(t)\rangle = H(\hat{p},\hat{q})|\psi(t)\rangle \tag{1.4}$$

where $\hat{H}$ is a Hamiltonian operator obtained from the classical Hamiltonian (1.2) by promoting the classical variable to the quantum operator variables. In this procedure there exists an ambiguity if p and q appear in a product form since $\hat{p}$ and $\hat{q}$ do not commute in quantum mechanics. This is called the ordering ambiguity. In this case one must define the quantum mechanics by specifying the order of the operators. Accordingly, to a classical system many quantum mechanical systems correspond.

A formal solution of Schrödinger equation (1.4) is given by:

$$|\psi(t)> = \underbrace{e^{i\hat{H}t}}_{\substack{\parallel \\ \text{evolution operator}}} |\psi(0)> \tag{1.5}$$

Using the coordinate representation

$$\hat{q}|q> = |q> q, \tag{1.6}$$

one defines the Feynman kernel

$$< q| e^{-i\hat{H}t} |q'> \tag{1.7}$$

which is a transition amplitude of a particle in q' at t=0 to q at later time t.

The extension of this one variable quantum mechanics to many variables is formally done by adding an appropriate index to the canonical variables:

$$q_a , p_a ; [\hat{q}_a, \hat{p}_b] = i\delta_{ab} \tag{1.8}$$

Field theories are the systems of many variables. The dynamical variables are on each space point. For example, for a scalar field theory one denotes the canonical coordinates and its conjugate momentum by $\phi(\vec{x})$, and $\pi(\vec{x})$ respectively. The canonical commutation is given by

$$[\hat{\phi}(\vec{x}), \hat{\pi}(\vec{x}')] = i\,\delta(\vec{x}-\vec{x}') \tag{1.9}$$

The coordinate representation in field theory is the $\phi(\vec{x})$-diagonal representation:

$$\hat{\phi}(\vec{x})\,|\phi> = |\phi>\phi(\vec{x}) \tag{1.10}$$

The wave function in this case is a functional of $\phi(\vec{x})$:

$$<\phi|\psi(t)> = \psi\,[\phi\,(\cdot), t] \tag{1.11}$$

The Feynman kernel is similarly defined:

$$<\phi|\; e^{-i\hat{H}t}\;|\phi'> \tag{1.12}$$

Path-Integral Quantization:

The path-integral quantization is defined by specifying the Feynman kernel as a sum over the paths i.e. the path-integral. Formally it is expressed as:

$$\begin{aligned} &<q|\; e^{-i\hat{H}t}\;|q'> \\ &= \int..\int Dq \;\; e^{\,i\int_0^t L(q(t'),\, \dot{q}(t'))dt'} \\ &q(t) = q \\ &q(0) = q' \end{aligned} \tag{1.13}$$

The precise definition of the path-integral is as following. We divide the time interval into N small segments:

$$t > t_{N-1} > t_{N-2} > \ldots . > t_3 > t_2 > t_1 > 0 \qquad (1.14)$$
$$\underset{t_N}{\parallel} \qquad\qquad\qquad\qquad \underset{t_0}{\parallel}$$

Then consider a set of N-1 variables, which we write $q(t_i)$ $(i=1,2,\ldots, N-1)$. We set $q(t_0) = q'$ and $q(t_N) = q$. Then, to a set of values of these variable a zigzag path from q' to q corresponds as shown in the following figure:

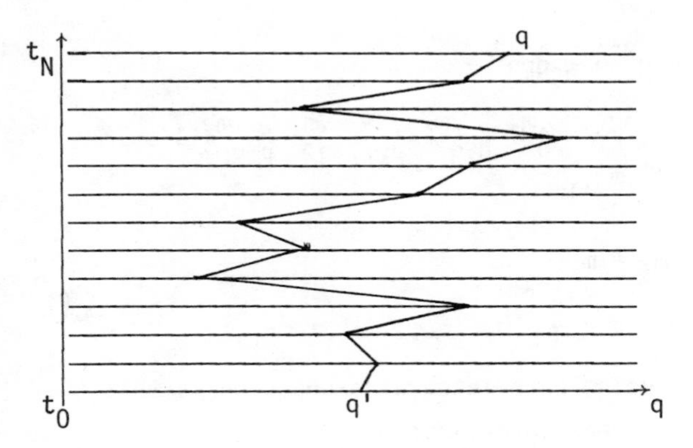

The action in (1.13) is replaced by

$$\int_0^t dt' \; L(q(t'), q(t')) \Rightarrow \sum_{i=1}^{N-1} (t_{i+1}-t_i)$$

$$L\left(\frac{q(t_{i+1}) + q(t_i)}{2} \quad , \quad \frac{q(t_{i+1}) - q(t_i)}{t_{i+1} - t_i}\right) \qquad (1.15)$$

and the integration measure by

$$Dq \Rightarrow \prod_{i=1}^{N-1} dq(t_i) \left(\frac{2\pi i}{t_{i+1}-t_i} \right)^{\frac{1}{2}} \qquad (1.16)$$

The path-integral (1.13) is the $N=\infty$ limit of this multiple integral.

This prescription is called as the mid-point prescription since in (1.15) $q(t')$ in the Lagrangian is replaced by the mid-point of $q(t_{i+1})$ and $q(t_i)$.

It is possible to obtain this path-integral expression from the operator formalism. The corresponding operator ordering to the mid-point prescription is the Weyl ordering defined by

$$(\hat{p}\hat{q})_W = \frac{1}{2}(\hat{p}\hat{q} + \hat{q}\hat{p})$$

$$(\hat{p}\hat{q}^3)_W = \frac{1}{4}(\hat{q}^3\hat{p} + \hat{q}^2\hat{p}\hat{q} + \hat{q}\hat{p}\hat{q}^2 + \hat{p}\hat{q}^3)$$

$$(\hat{p}^\ell\hat{q}^m)_W = \frac{1}{A}\sum_{\substack{m_1 m_2 \ldots m_{\ell+1} \\ \sum_i^{\ell+1} m_i = m}} \hat{q}^{m_1}\,\hat{p}\,\hat{q}^{m_2}\,\hat{p}\ldots\hat{q}^{m_\ell}\,\hat{p}\,\hat{q}^{m_{\ell+1}}$$

$$A = \sum_{m_1 m_2 \ldots m_\ell +1} \delta_{m\,,\,\sum_i^{\ell+1} m_i} \tag{1.17}$$

The advantage of using path-integral expression is that the expression is formally Lorenz covariant:

$$\langle \phi |\, e^{-i\hat{H}t}\, | \phi' \rangle = \int \ldots \int D\phi\; e^{i\int_0^t d\vec{x}\, \mathscr{L}(\phi(x),\, \partial_\mu \phi(x))} \tag{1.18}$$

$$\phi(\vec{x},t) = \phi(\vec{x})$$
$$\phi(\vec{x},0) = \phi'(x)$$

Euclidean Path-Integral:

In quantum statistical mechanics one considers the partition function defined by

$$Z = \mathrm{tr}\, e^{-\beta\hat{H}} = \int dq \,\langle q |\, e^{-\beta\hat{H}}\, | q \rangle\,. \tag{1.19}$$

So, the relevant matrix element to consider is

$$< q| \, e^{-\beta\hat{H}} \, |q'> \tag{1.20}$$

which can be obtained from the Feynman kernel (1.7) by replacing t by $-i\beta$. This is accomplished by considering a complex t-plane and by rotating the time axis by 90^0 clockwise to the negative imaginary axis (Wick rotation).

The action integral is then

$$i \int L \, dt \; \rightarrow \; -S \tag{1.21}$$

where

$$S = - \int_0^\beta d\tau \; L(q, i\dot{q}) \tag{1.22}$$

which we call Euclidean action.

For example, for the following Minkowsky Lagrangian

$$L = \frac{1}{2} \dot{q}^2 - V(q) \tag{1.23}$$

one obtains

$$S = \int \; d\tau \; [\; \frac{1}{2} \dot{q}^2 + V(q) \;] \tag{1.24}$$

For a field theory in d-dimensional space

$$L = \int d\vec{x} \; \{ \frac{1}{2} \dot{\phi}^2 - \underbrace{(\; \frac{1}{2}(\vec{\nabla}\phi)^2 + \frac{1}{2} m^2\phi^2 + \ldots .)}_{\substack{|||\\ V}} \} \tag{1.25}$$

one obtains

$$S[\phi] \equiv S = \int d\vec{x} \int_0^\beta dx_0 \; \{ \frac{1}{2}(\partial_0\phi)^2 + \frac{1}{2}(\vec{\nabla}\phi)^2 + \frac{1}{2} m^2\phi^2 + \ldots \} \tag{1.26}$$

The partition function (1.19) is then given by

$$Z = \int \ldots \int D\phi \; e^{-S[\phi]}$$
$$\phi(\vec{x},0) = \phi(\vec{x},\beta) \tag{1.27}$$

which has a form of partition function of classical statistical mechanics in d + 1 dimensions. Thus, "d-dimensional quantum statistical mechanics = d + 1 dimensional classical statistical mechanics".

Parisi-Wu Stochastic Quantization:

The stochastic quantization of Parisi and Wu is a method of quantization for Bose fields in Euclidean space time. The correlation functions of scalar field defined by the standard Euclidean path integral are computed by the following prescription:

$$\begin{aligned} &< \phi(x_1)\phi(x_2)\ldots\ldots\phi(x_n) > \equiv \\ &\equiv \int \ldots \int D\phi \;\; \phi(x_1)\phi_\eta(x_2)\ldots\ldots\phi(x_n)\; e^{-S[\phi]} / \int .. \int D\phi \; e^{-S[\phi]} \\ &= \lim_{\tau\to\infty} < \phi_\eta(x_1)\phi_\eta(x_2)\ldots\ldots\phi_\eta(x_n,\tau) >_\eta \end{aligned} \tag{1.28}$$

where $\phi_\eta(x,\tau)$ is a solution of Langevin equation

$$\frac{\partial}{\partial\tau}\phi_\eta(x,\tau) = - \left.\frac{\delta S[\phi]}{\partial\phi(x)}\right|_{\phi(x) = \phi_\eta(x,\tau)} + \eta(x,\tau) \tag{1.29}$$

with a given initial condition

$$\phi_\eta(x,0) = \phi_0(x) \tag{1.30}$$

and $< \ldots\ldots\ldots >_\eta$ is a white noise average:

$$< \eta(x,\tau) >_\eta = 0$$

$$< \eta(x,\tau)\ \eta(x',\ \tau') >_\eta = 2\delta(x-x')\delta(\tau-\tau')$$

(1.31)

In order to avoid some mathematical inconsistency we implicitly assume that the η average (4) is defined by the following limit:

$$< \ldots\ldots >_\eta =$$

$$= \lim_{\Lambda\to\infty} \frac{\int\cdots\int D\eta \ldots\ldots e^{\frac{-1}{4}\int dx d\tau d\tau' a_\Lambda(\tau-\tau')\ \eta(x,\tau)\ \eta(x,\tau')}}{\int\cdots\int D\eta\ e^{\frac{-1}{4}\int dx d\tau d\tau' a_\Lambda(\tau-\tau')\ \eta(x,\tau)\ \eta(x',\ ')}}$$

(1.32)

where $a_\Lambda(\tau-\tau')$ is a symmetric regulator function such that

$$a_\Lambda(\tau) = a_\Lambda(-\tau)$$

$$\int d\tau'\ a_\Lambda(\tau-\tau') = 1$$

$$\lim_{\Lambda\to\infty} a_\Lambda(\tau-\tau') = \delta(\tau-\tau')$$

(1.33)

In order to show how this quantization works, let us consider a simple example of a free scalar field. The Langevin equation is given by

$$\frac{\partial}{\partial\tau}\phi(x,\tau) = (\Box - m^2)\phi(x,\tau) + \eta(x,\tau) \qquad (1.34)$$

The solution $\phi_\eta(x,\tau)$ with the initial condition $\phi_\eta(x,0) = 0$ is given by

$$\tilde{\phi}_\eta (k,\tau) = \int_0^\tau d\tau' \, e^{(\tau'-\tau)(k^2+m^2)} \, \tilde{\eta}(k,\tau') \tag{1.35}$$

where $\tilde{\phi}_\eta(k,\tau)$ and $\tilde{\eta}(k,\tau)$ are Fourier transform, accordingly from (1.31)

$$< \tilde{\eta}(k,\tau) \, \tilde{\eta}(k',\tau') >_\eta = 2\delta(k+k') \, \delta(\tau-\tau') \tag{1.36}$$

Thus,

$$< \tilde{\phi}(k,\tau) \, \tilde{\phi}(k',\tau') >_\eta =$$

$$= \delta(k+k') \frac{1}{k^2+m^2} \left(e^{|\tau-\tau'| \, (k^2+m^2)} - e^{-|\tau+\tau'| \, (k^2+m^2)}\right) \tag{1.37}$$

which becomes the standard Feynman propagator in the limit $\tau=\tau' \to \infty$.

For the interacting field such as ϕ^4 theory the Langevin equation has an additional term, $-g\phi^3$. The perturbative solution is expressed by tree graphs:

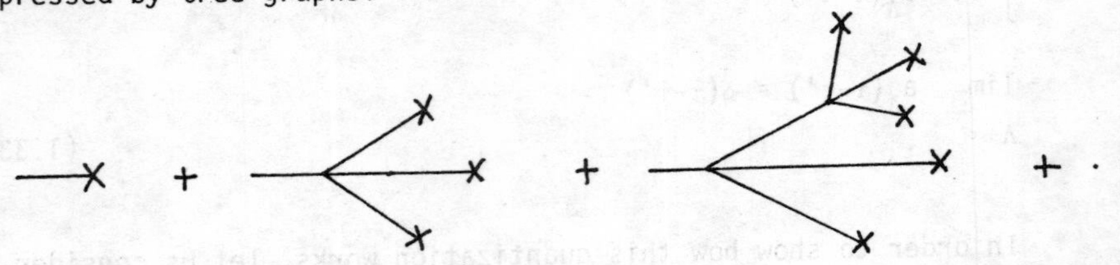

where x represents the random source η. The η average connects a pair of crosses as

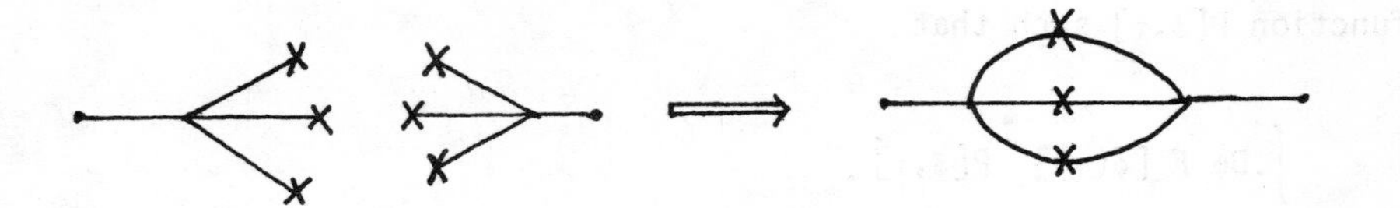

η - average

Thus, one obtains a similar expression to the standard Feynman diagram.

The stochastic quantization is based on the Langevin equation (1.29), which is formally obtained by considering the field as a function of a fictitious time τ as well as the ordinary space time x and by adding $\dot{\phi}$ and a random source term η to the classical equation $\frac{\delta S[\phi]}{\delta\phi(x)} = 0$. In this sense the quantization is carried out once the classical equation of motion is known. This feature contrasts with other quantization methods, the canonical quantization based on the Hamiltonian formalism and the path-integral quantization by means of the Lagrangian.

To what extent is the stochastic quantization method equivalent to the other methods? What are the advantages of using this method? Focussing these points in this lecture I will review the recent developments.

II. GENERAL THEORY[2]

In this section we formulate a general theory of stochastic processes in such a way to serve for the extension of the stochastic quantization described in the previous section to a wider class of field theories.

We assume first the existence of probability distribution

function $P[\phi,\tau]$ such that

$$\int D\phi \, F[\phi(\cdot)] \; P[\phi,\tau]$$

$$= \langle F[\phi_\eta(\cdot,\tau)] \rangle_\eta \tag{2.1}$$

where $F[\phi(\cdot)]$ is an arbitrary functional of $\phi(x)$. $\phi_\eta(x,\tau)$ is a solution of Langevin equation which is yet to be determined. In order to be the stochastic quantization (1.1) we demand

$$\lim_{\tau\to\infty} P[\phi,\tau] = e^{-S[\phi]} \Big/ \int \cdots \int D\phi \, e^{-S[\phi]} \tag{2.2}$$

and

$$\int D\phi \, P[\phi,\tau] = 1 \tag{2.3}$$

(2.3) is merely the statement $P[\phi,\tau]$ being a probability distribution.

Next we assume that the time development of $P[\phi,\tau]$ is governed by a Schrödinger-type equation, the Fokker-Planck equation:

$$-\frac{\partial}{\partial\tau} P[\phi,\tau] = H_{FP} \, P[\phi,\tau] \tag{2.4}$$

We then ask what are the possible forms of H_{FP} which leads to (2.2) and (2.3), and what are the corresponding Langevin equations which assure the relation (2.1).

It is advantageous to use Dirac's canonical operator formalism[3]:

$$[\hat{\pi}(x), \hat{\phi}(x')] = -i\delta(x-x') \tag{2.5}$$

$$\hat{\phi}(x)\ |\phi> = \ |\phi>\ \phi(x) \tag{2.6}$$

$$P[\phi,\tau] \equiv <\phi|\ P(\tau)> \tag{2.7}$$

$$<\phi|\ \hat{\pi}(x)\ |\ P(\tau)> = -\ i\ \frac{\delta}{\delta\phi(x)}\ <\phi|\ P(\tau)> \tag{2.8}$$

The Fokker-Planck equation is then

$$-\frac{\partial}{\partial\tau}\ |\ P(\tau)> = \hat{H}_{FP}\ |P(\tau)> \tag{2.9}$$

A formal solution of $|P(\tau)>$ is given by

$$|P(\tau)> = e^{-\hat{H}_{FP}\tau}\ |P(0)> \tag{2.10}$$

Integrating (2.8) functionally over ϕ we obtain

$$\int D\phi\ <\phi|\ \hat{\pi}(x)\ |P(\tau)> = 0$$

for any $|P(\tau)>$. Namely,

$$\int D\phi\ |\phi> \equiv |0>\ ,\quad \hat{\pi}(x)\ |0> = 0 \tag{2.11}$$

With this notation (2.1) and (2.3) are expressed as

$$<0|\ F[\hat{\phi}(\cdot)]\ |P(\tau)> = <F[\phi_\eta(\cdot,\tau)]>_\eta \tag{2.12}$$

$$<0|\ P(\tau)> = 1 \tag{2.13}$$

Taking time derivative of (2.13) and using (2.9) we obtain

$$< 0|\, \hat{H}_{FP}\, |P(\tau)> = 0 \tag{2.14}$$

This is satisfied due to (2.11) if $\hat{H}_{FP}$ contains $\hat{\pi}(x)$ on its left.

Let E_n be an eigenvalue of $\hat{H}_{FP}$, and $|\chi_n>$ and $<\xi_n|$ be the corresponding right and left eigenvector respectively.

$$\hat{H}_{FP}\, |\chi_n> \;=\; E_n\, |\chi_n> \quad , \quad <\xi_n|\, \hat{H}_{FP} = E_n\, <\xi_n|$$

$$<\xi_n|\chi_m> \;=\; \delta_{nm}$$

$$\sum_n |\chi_n><\xi_n| = 1 \tag{2.15}$$

Using (2.10) we obtain

$$P[\Phi,\tau] = \sum_n e^{-E_n\tau}\, \chi_n[\phi]\, <\xi_n|\, P(0)> \tag{2.16}$$

If

$$R_e\, E_n > 0 \qquad (n \neq 0) \tag{2.17}$$

and

$$E_0 = 0$$

namely, there is an energy gap above the ground state, and further the ground state is non-degenerate, we obtain

$$\lim_{\tau\to\infty} P[\phi,\tau] = \chi_0[\phi] < \xi_0 | P(0) > \tag{2.18}$$

Comparing (2.2) and (2.18) we require

$$\chi_0[\phi] \quad < \xi_0 | \; P(0) > \; = \; e^{-S[\phi]} \Big/ \int D\phi \; e^{-S[\phi]} \tag{2.19}$$

We note

$$e^{-S[\phi]} = < \phi | \; e^{-S[\hat{\phi}]} \; | 0 > \tag{2.20}$$

Thus

$$| \chi_0 > \; \propto e^{-S[\hat{\phi}]} \; | 0 > \tag{2.21}$$

Therefore, using (2.11) we obtain

$$(\hat{\pi}(x) + [\hat{\pi}(x), S[\hat{\phi}]]) \quad | \chi_0 > \; =$$

$$= (\hat{\pi}(x) - i \frac{\delta S[\hat{\phi}]}{\delta \hat{\phi}(x)}) \quad | \chi_0 > \; = 0 \tag{2.22}$$

This expression and $\hat{H}_{FP} | \chi_0 > = 0$ is compatible if $\hat{H}_{FP}$ has $\hat{\pi}(x) + [\pi(x), S[\hat{\phi}]]$ on its right. Thus, we obtain the following general form for $\hat{H}_{FP}$

$$\hat{H}_{FP} = \iint dx dx' \; \hat{\pi}(x) \; \hat{K}(x,x') \; (\hat{\pi}(x') + [\hat{\pi}(x'), S[\hat{\phi}]]) \tag{2.23}$$

where $\hat{K}(x,x')$ is in general a functional of $\hat{\phi}(x)$ and $\hat{\pi}(x)$. If we can choose $\hat{K}(x,x')$ such that the real part of eigenvalues of $\hat{H}_{FP}$ is positive definite, then (2.23) is a possible form.

For the case of Hermitean, $S^{+} = S$, we choose $\hat{K}(x,x') = \delta(x-x')$

and make a similarity transformation to obtain the following positive definite form:

$$e^{\frac{1}{2}\hat{S}}\ \hat{H}_{FP}\ e^{-\frac{1}{2}\hat{S}} = \int dx\ \hat{Q}(x)\ \hat{Q}(x) \tag{2.24}$$

$$\hat{Q}(x) = \hat{\pi}(x) + \frac{1}{2}\,[\hat{\pi}(x),\ \hat{S}] \tag{2.25}$$

Even for non-Hermitian S, if we can choose $\hat{K}(x,x')$ such that after making an appropriate similarity transformation $\hat{H}_{FP}$ becomes a positive form, then we can use (2.23) to derive the corresponding Langevin equation. However, whether this can done for a general S or not is not known yet.

Using the identity (2.12) we like to obtain the equation that $\phi_\eta(x,\tau)$ should satisfy. We restrict here to the case of $K(x,x') = \delta(x-x')$, but the extension to the general case is straightforward.

First define the following Heisenberg operator

$$\hat{\phi}(x,\tau) = e^{\hat{H}_{FP}\tau}\ \hat{\phi}(x)\ e^{-\hat{H}_{FP}\tau} \tag{2.26}$$

then

$$< 0|\ F[\hat{\phi}(\cdot)]\ |P(\tau)> \ = \ < 0|\ F[\hat{\phi}(\cdot,\tau)]\ \ |P(0)> \tag{2.27}$$

Taking time derivative we obtain

$$\frac{\partial}{\partial\tau} < 0|\ F[\hat{\phi}(\cdot,\tau)]\ |P(0)>$$

$$= < 0| \int \left(\frac{\delta\hat{F}}{\delta\phi(x,\tau)}\ \dot{\hat{\phi}}(x,\tau)\right)_W dx\ |P(0)> \tag{2.28}$$

where $(\quad)_W$ is the Weyl ordering defined in (1.17).

Since $\hat{\phi}(x,\tau)$ is defined by (2.26) and since $\hat{H}_{FP}$ is given by (2.23), we obtain

$$\dot{\hat{\phi}}(x,\tau) = -i[\hat{\pi}(x,\tau), \hat{S}] - 2i\hat{\pi}(x,\tau) \tag{2.29}$$

It is not difficult to prove

$$-2i < 0| \int (\frac{\delta \hat{F}}{\delta\phi(x,\tau)} \hat{\pi}(\vec{x},\tau))_W dx |P(0)>$$

$$= <0| \int \frac{\delta^2 \hat{F}}{\delta\phi^2(x,\tau)} dx |P(0)> \tag{2.30}$$

Inserting (2.29) into (2.28) and using (2.30) we obtain

$$\frac{\partial}{\partial\tau} <0| F[\phi(\cdot,\tau)] |P(0)> =$$

$$= -i <0| \int \frac{\delta\hat{F}}{\delta\phi(x,\tau)} [\hat{\pi}(x,\tau), S] dx |P(0)> +$$

$$+ <0| \int dx \frac{\delta^2 F}{\delta\phi^2(x,\tau)} |P(0)> \tag{2.31}$$

We note that in the right hand side of (2.31) only a product of $\hat{\phi}$ appears. Thus we can use the identity (2.12). We obtain

$$\frac{\partial}{\partial\tau} < F[\phi_\eta(\cdot,\tau)] >_\eta =$$

$$= -<\int \frac{\delta F}{\delta\phi_\eta(x,\tau)} \frac{\delta S}{\delta\phi_\eta(x,\tau)} >_\eta + <\frac{\delta^2 F}{\delta\phi_\eta^2(x,\tau)} dx >_\eta \tag{2.32}$$

Since

$$\frac{\partial}{\partial\tau} < F[\phi_\eta(\cdot,\tau)] >_\eta = \int dx < \frac{\delta F}{\delta\phi_\eta(x,\tau)} \frac{\partial}{\partial\tau} \phi_\eta(x,\tau) >_\eta \tag{2.33}$$

and since we shall prove

$$< \int \frac{\delta^2 F}{\delta\phi^2(x,\tau)} dx >_\eta = < \int dx \frac{\delta F}{\delta\phi(x,\tau)} \eta(x,\tau) >_\eta \tag{2.34}$$

we obtain the Langevin equation:

$$\frac{\partial\phi_\eta(x,\tau)}{\partial\tau} = - \frac{\delta S[\phi_\eta(\cdot,\tau)]}{\delta\phi_\eta(x,\tau)} + \eta(x,\tau) \tag{2.35}$$

Proof of (2.34):

$$< \int dx \frac{\delta F}{\delta\phi(x,\tau)} \eta(x,\tau) >_\eta = \lim_{\Lambda\to\infty} \int d\tau' \, a_\Lambda^{-1}(\tau-\tau') \times$$

$$\times < \int dx \frac{\delta}{\delta\eta(x,\tau')} \frac{\delta F}{\delta\phi_\eta(x,\tau)} >_\eta$$

$$= \lim_{\Lambda\to\infty} \int d\tau' \, 2\, a_\Lambda^{-1}(\tau-\tau') < \int dxdy \frac{\delta\phi_\eta(y,\tau)}{\delta\,\eta(x,\tau')} \frac{\delta^2 F}{\delta\phi_\eta(x,\tau)\delta\phi_\eta(y,\tau)} >_\eta \tag{2.36}$$

where a_Λ^{-1} is an inverse function of a_Λ:

$$\int a_\Lambda^{-1}(\tau-\tau') \, a_\Lambda(\tau'-\tau'') \, d\tau' = \delta(\tau-\tau'') \tag{2.37}$$

Next we prove

$$\lim_{\Lambda\to\infty} \int a_\Lambda^{-1}(\tau-\tau') \, d\tau' \frac{\delta\phi_\eta(y,\tau)}{\delta\eta(x,\tau')} = \frac{1}{2}\,\delta(x-y) \tag{2.38}$$

then (2.34) follows. Integrating langevin equation we obtain

$$\phi_\eta(x,\tau) = \phi_\eta(x,0) - \int_0^\tau d\tau'' \frac{\delta S[\phi_\eta(\cdot,\tau'')]}{\delta\phi_\eta(x,\tau'')} + \int_0^\tau d\tau' \; \eta(x,\tau') \tag{2.39}$$

Thus,

$$\frac{\delta\phi_\eta(x,\tau)}{\delta\eta(y,\tau)} = \theta(\tau-\tau') - \int_0^\tau d\tau \int \frac{\delta\phi_\eta(y,\tau'')}{\delta\eta(x,\tau)} \; \frac{\delta^2 S}{\delta\phi_\eta(x,\tau'')\delta\phi_\eta(y,\tau')} \tag{2.40}$$

Inserting it into (2.38) and using

$$\lim_{\Lambda\to\infty} \int_0^\tau a_\Lambda^{-1}(\tau-\tau') \; \theta(\tau-\tau') \; d\tau' = \frac{1}{2}$$

$$\lim_{\Lambda\to\infty} \int^\tau d\tau' \int_0^\tau d\tau'' \; a_\Lambda^{-1}(\tau-\tau') \; \theta(\tau''-\tau') = 0 \tag{2.41}$$

we obtain (2.38).

III. GENERALIZATIONS

In this section we generalize the stochastic quantization method to include U(N) lattice gauge theory and Fermi fields.

General Fokker-Planck Hamiltonian and Corresponding Langevin Equation[2,4]:

We consider a special case of the Hamiltonian (2.23) in which K(x,x') depends only on $\phi(x)$ and factorizable:

$$K(x,x';\phi) = \int dz \; G(x,z;\phi) \; G(y,z;\phi) \tag{3.1}$$

In this case if S is real one can prove that eigenvalues of $\hat{H}_{FP}$ are semi-positive definite. Using the method of previous section we obtain the following Langevin equation:

$$\frac{\partial}{\partial\tau}\phi_\eta(x,\tau) = -\int dy\, K(x,y;\phi_\eta)\frac{\delta S}{\partial\phi_\eta(y,\tau)} + $$

$$+ \int dydz\, G(x,z;\phi_\eta)\frac{\delta G}{\delta\phi_\eta(y,\tau)} + \int dy\, G(x,y;\phi_\eta)\,\eta(y,\tau) \tag{3.2}$$

This is a generalized Langevin equation by which one obtains the same correlation functions as previous case (a special case of $G(x,y;\phi) = \delta(x-y)$) in the $\tau\to\infty$ limit. Since G is arbitrary one can choose it as one's convenience.

Stochastic Quantization of U(N) Lattice Gauge Theory[5]:

In Wilson's lattice gauge theory one considers an SU(N) matrix on each link of square lattice. A link is specified by x and μ, where x is the position of a lattice point and μ is the direction along the link. Let $U_\mu(x)$ be an NxN unitary matrix on the link $x\mu$ and $dU_\mu(x)$ be the Haar measure. The partition function of the model is defined by the following path-integral:

$$Z = \int \prod_{x,\mu} dU_\mu(x)\, e^{-S[U]} \tag{3.3}$$

where S[U] is given by

$$S[U] = \frac{1}{4g^2}\sum_{x,\mu,\nu}[(1 - tr(U_\mu(x)\, U_\nu(x+a\hat{\mu})\, U^\dagger_\mu(x+a\hat{\nu})\, U^\dagger_\nu(x))) + h.c.] \tag{3.4}$$

In this expression $x + a\hat{\mu}$ refer to the lattice point next to x in the μ direction.

In (3.3), the integration is in a product of U(N) group spaces, on the other hand in our previous Eculidean path-integral (1.27) the integration was in a product of real open spaces. Thus, appropriately modifying the derivative by Lie derivative we still use the general theory of the previous section.

$$D\phi \;\rightarrow\; \Pi dU_\mu(x)$$

$$\pi(\vec{x}) = -\, i\,\frac{\delta}{\delta\phi(x)} \;\rightarrow\; E^\alpha_\mu \tag{3.5}$$

$E^\alpha_\mu(x)$ is a Lie derivative of matrix $U_\mu(x)$, namely

$$E^\alpha_\mu(x)\,(U_{\mu'}(x'))_{ij} = \delta_{xx'}\,\delta_{\mu\mu'}\,(t^\alpha\,U_\mu(x))_{ij} \tag{3.6}$$

where t^α is a fundamental representation of U(N) Lie algebra. We normalize t^α such that

$$tr(t^\alpha t^\beta) = \delta_{\alpha\beta}\,, \quad \sum_\alpha (t^\alpha)_{ij}\,(t^\alpha)_{k\ell} = \delta_{i\ell}\,\delta_{jk}$$

$$[t^\alpha, t^\beta] = \sum_\gamma C^{\alpha\beta}_\gamma\, t^\gamma \tag{3.7}$$

(Caution: Our normalization is different from the standard one by factor $\frac{1}{\sqrt{2}}$, accordingly the structure constant $C^{\alpha\beta}_\gamma$ also.)

The Fokker-Planck Hamiltonian we use is

$$H_{FP} = \sum_{x\mu} E^\alpha_\mu(x)\,(E^\alpha_\mu(x) + (E^\alpha_\mu(x)\,S[U])) \tag{3.8}$$

which corresponds to (2.23) with K = 1. It is straightforward by using the method of previous section to obtain the following Langevin equation:

$$- i(\frac{\partial}{\partial\tau} U_\mu(x,\tau))\, U_\mu^\dagger(x,\tau)$$

$$= - i(E_\mu(x)\, S[U(\cdot,\tau)]) + \eta_\mu(x,\tau) \tag{3.9}$$

where

$$E_\mu = \sum_\alpha E_\mu^\alpha\, t^\alpha\ ,\ \eta_\mu = \sum_\alpha \eta_\mu^\alpha\, t^\alpha \tag{3.10}$$

and

$$< \eta_\mu^\alpha(x,\tau)\ \eta_\nu^\beta\ (x',\tau')\ >_\eta\ = 2\ \delta_{\mu\mu'}\ \delta_{xx'}\ \delta_{\alpha\alpha'}\ \delta(\tau-\tau') \tag{3.11}$$

The Langevin equation (3.9) is used in the talk given by Dr. A. Guha.

Stochastic Quantization of Fermi Fields[6]:

The path-integral expression of Fermi fields to be considered is given by

$$\int D\psi D\bar{\psi}(.....)\ e^{-S[\psi,\bar{\psi}]} / \int D\psi D\bar{\psi}\ e^{-S[\psi,\bar{\psi}]} \tag{3.12}$$

where ψ and $\bar{\psi}$ are independent Grassmann variables. We restrict S to be bi-linear form, since in the most of interesting theories the Fermions appear bi-linearly in the Lagrangian and since even if Fermions appear in quartic form one can bi-linearize it by introducing a set of auxiliary fields:

$$S = \int \bar{\psi}(x)\ G\ \psi(x)\ dx \tag{3.13}$$

G may contain not only the derivative operators but also other fields, which we treat here as external fields. In general,

$$G^\dagger \neq G \tag{3.14}$$

The Fokker Planck Hamiltonian to be used is

$$H_{FP} = \int dx \, [\frac{\delta}{\delta\psi(x)} \; G^\dagger \; (\frac{\delta}{\delta\bar{\psi}(x)} + \frac{\delta S}{\delta\bar{\psi}(x)}) -$$

$$- \frac{\delta}{\delta\bar{\psi}(x)} \; G^{T\dagger} \; (\frac{\delta}{\delta\psi(x)} + \frac{\delta S}{\delta\psi(x)})]$$

$$= \int dx \, [\frac{\delta}{\delta\psi(x)} \; G^\dagger \; (\frac{\delta}{\delta\bar{\psi}(x)} + G \, \psi(x)) -$$

$$- \frac{\delta}{\delta\bar{\psi}} \; G^{T\dagger} \; (\frac{\delta}{\delta\psi(x)} - G^T\bar{\psi}(x))] \tag{3.15}$$

This was constructed based on the general discussion of section II. The arbitrary function and sign are fixed in such a way H_{FP} has semi-positive definite eigenvalues, which we shall prove later.

Following the general discussion of II, we derive the corresponding Langevin equation:

$$\frac{\partial}{\partial\tau} \psi(x,\tau) = - G^\dagger \, G \, \psi(x,\tau) + \frac{1}{2} G^\dagger\eta_1 + \eta_2$$

$$\frac{\partial}{\partial\tau} \bar{\psi}(x,\tau) = - (GG^\dagger)^T \, \bar{\psi}(x,\tau) + \bar{\eta}_1 + \frac{1}{2} G^{\dagger T}\bar{\eta}_2 \tag{3.16}$$

$$< \eta_\alpha(x,\tau) \, \bar{\eta}_\beta(x',\tau') >_\eta = - < \bar{\eta}_\beta(x',\tau') \, \eta_\alpha(x,\tau) >_\eta$$

$$= 2 \, \delta_{\alpha\beta} \, \delta(x-x') \, a_\Lambda(\tau-\tau'). \tag{3.17}$$

Next we prove the positivity of H_{FP}. The method we use is based on the work of Fukai et al[7]. Let us consider two sets of Fermi operators:

$$\{\hat{\psi}, \hat{\psi}^{\dagger}\} = 1 \quad , \quad \{\hat{\bar{\psi}}, \hat{\bar{\psi}}^{\dagger}\} = 1$$

$$\{\hat{\psi}, \hat{\bar{\psi}}^{\dagger}\} = 0, \text{ etc.} \tag{3.18}$$

The coherent state representation defined by

$$|\psi,\bar{\psi}> \;=\; |0,0> + \psi|1,0> + \bar{\psi}|0,1> + \psi\bar{\psi}|1,1> \tag{3.19}$$

represents $\hat{\psi}$ and $\hat{\psi}^{\dagger}$ as follows:

$$\hat{\psi}^{\dagger}|\psi,\bar{\psi}> \;=\; \frac{\delta}{\delta\psi}|\psi,\bar{\psi}> \;, \quad \hat{\bar{\psi}}^{\dagger}|\psi,\bar{\psi}> \;=\; \frac{\delta}{\delta\bar{\psi}}|\psi,\bar{\psi}>$$

$$\hat{\psi}\,|\psi,\bar{\psi}> \;=\; \psi\,|\psi,\bar{\psi}> \quad , \quad \hat{\bar{\psi}}\,|\psi,\bar{\psi}> \;=\; \bar{\psi}\,|\psi,\bar{\psi}> \tag{3.20}$$

$|0,0>$, $|0,1>$ etc. in (3.19) are the Fermion number representation of $\hat{\psi}^{\dagger}\hat{\psi}$ and $\hat{\bar{\psi}}^{\dagger}\hat{\bar{\psi}}$.

The Fokker Planck Hamiltonian (3.15) can be considered as a coherent state representation of the following operator Hamiltonian.

$$H_{FP} = \int dx\,[\hat{\psi}^{\dagger}\,G^{\dagger}\,(\hat{\bar{\psi}}^{\dagger} + G\,\hat{\psi}) - \hat{\bar{\psi}}\;G^{T\dagger}\,(\hat{\psi}^{\dagger} - G^{T}\hat{\bar{\psi}})] \tag{3.21}$$

By the following similarity transformation

$$\hat{H}_{FP} = e^{-\int dx\,\hat{\psi}^{\dagger}\,G^{-1}\hat{\bar{\psi}}^{\dagger}}\;H_{FP}\;e^{+\int dx\,\hat{\psi}^{\dagger}\,G^{-1}\,\hat{\bar{\psi}}^{\dagger}} \tag{3.22}$$

we obtain the positive definite form:

$$\hat{H}_{FP} = \int dx\,(\hat{\psi}^{\dagger}\,G^{\dagger}G\,\hat{\psi} + \hat{\bar{\psi}}^{\dagger}(GG^{\dagger})^{T}\hat{\bar{\psi}}) \tag{3.23}$$

IV. PATH-INTEGRAL FORMULATION OF STOCHASTIC PROCESSES AND HIDDEN SUPERSYMMETRY[8]

Let us define a generating functional

$$Z[j] = \langle \exp \int dx\, j(x)\, \phi_\eta(x,\tau) \rangle_\eta \tag{4.1}$$

and then convert it to path-integral form by using Faddeev-Popov technique. (Since x is a spectator variable we omit it in the following expression.)

$$Z[j] = \int \cdots \int D\phi\, \underbrace{\delta(\phi - \phi_\eta)}_{\overbrace{\delta(\dot{\phi} + \frac{\partial S}{\partial \phi} - \eta) \left|\left|\frac{\delta\eta}{\delta\phi}\right|\right|}}\, e^{ij\phi(\tau)}\, D\eta\, e^{-\frac{1}{4}\int_0^\tau d\tau\, \eta^2(\tau)} \tag{4.2}$$

The Jacobian with the retarded boundary condition (1.30) can be computed as

$$\left|\left|\frac{\delta\eta}{\delta\phi}\right|\right| = \exp\left(\frac{1}{2}\int_0^\tau d\tau\, \frac{\partial^2 S}{\partial\phi^2(\tau)}\right) \tag{4.3}$$

We shall give details of calculation later. Inserting this into (4.2) we obtain

$$Z[j] = \underset{\phi(0)=\phi_0}{\int \cdots \int} D\phi\, e^{j\phi(\tau)}\, e^{-\int_0^\tau d\tau\left[(\dot{\phi} + \frac{\partial S}{\partial\phi})^2 - \frac{1}{2}\frac{\partial^2 S}{\partial\phi^2}\right]}$$

$$= \int d\phi\, e^{j\phi - \frac{1}{2}(S[\phi] - S[\phi_0])} \underset{\substack{\phi(\tau)=\phi \\ \phi(0)=\phi_0}}{\int \cdots \int} D\phi\, e^{-\int_0^\tau d\tau\, L} \tag{4.4}$$

where L is Fokker Planck Lagrangian given by

$$L = \frac{1}{4}\dot{\phi}^2 + \frac{1}{4}\left(\frac{\partial S}{\partial \phi}\right)^2 - \frac{1}{2}\frac{\partial^2 S}{\partial \phi^2} \equiv \frac{1}{4}\dot{\phi}^2 + V[\phi] \tag{4.5}$$

Using the relation between the path-integral and the operator matrix element, the last factor of (4.4) can be written as

$$\int \cdots \int D\phi \; e^{-\int_0^\tau d\tau L} = \langle \phi | \, e^{-\tau \hat{H}} \, | \phi_0 \rangle \tag{4.6}$$

$$\phi(\tau) = \phi$$
$$\phi(0) = \phi_0$$

where $\hat{H}$ is the Fokker Planck Hamiltonian

$$\hat{H} = \hat{\pi}^2 + V[\phi] = \left(\hat{\pi} + \frac{i}{2}\frac{\partial S}{\partial \phi}\right)\left(\hat{\pi} - \frac{i}{2}\frac{\partial S}{\partial \phi}\right) \equiv \hat{Q}^\dagger \hat{Q} \tag{4.7}$$

which has semi-positive definite form. The zero energy state of $\hat{H}$ is given by

$$\hat{H}\,|0\rangle = 0, \quad \langle \phi | 0 \rangle = e^{-\frac{1}{2}S[\phi]} / \left(\int d\phi \, e^{-S}\right)^{\frac{1}{2}} \tag{4.8}$$

which we assume to be normalizable. (i.e. $\int d\phi \, e^{-S} < \infty$). Inserting (4.5) into (4.4) and expanding by a complete set of eigenfunctions of $\hat{H}$ we obtain

$$Z[j] = \int d\phi \; e^{j\phi - \frac{1}{2}S[\phi]} \sum_n e^{-E_n} \langle \phi | n \rangle \langle n | \phi_0 \rangle \, e^{\frac{1}{2}S[\phi_0]} \tag{4.9}$$

Since $E_n \geq 0$ for $n \neq 0$, we obtain

$$Z[j] \xrightarrow[\tau \to \infty]{} \int d\phi \; e^{j\phi} \, e^{-S[\phi]} / \left(\int d\phi \, e^{-S}\right)^{1/2} \tag{4.10}$$

Here we assumed a <u>non-degenerate</u> zero energy ground state. (4.1) and (4.10) consist of another non-perturbative proof of the stochastic quantization.

Calculation of Jacobian (4.3)

$$\left|\left|\frac{\delta\eta}{\delta\phi}\right|\right| = \det\left[\frac{\partial}{\partial\phi(\tau')}\left(\dot\phi(\tau) + \frac{\partial S}{\partial\phi(\tau)}\right)\right]$$

$$= \det\left[\left(\partial_\tau + \frac{\partial^2 S}{\partial\phi^2(\tau)}\right)\delta(\tau-\tau')\right] \equiv \det\left(<\tau|M|\tau'> \right)$$

$$\equiv \det M \tag{4.11}$$

where

$$M = \hat\partial_\tau + \frac{\partial^2 S}{\partial\phi^2(\hat\tau)} = \hat\partial_\tau\left(1 + \hat\partial_\tau^{-1}\frac{\partial^2 S}{\partial\phi^2(\hat\tau)}\right) \tag{4.12}$$

$$\ell n \det M = \mathrm{tr}\, \ell n\, M$$

$$= \mathrm{tr}\left[\ell n\, \hat\partial_\tau^{-1} + \ell n\left(1 + \hat\partial_\tau^{-1}\frac{\partial^2 S}{\partial\phi^2(\hat\tau)}\right)\right] \tag{4.13}$$

The first term gives merely a constant so we will omit it. The second term is

$$\int d\tau < \tau|\, \ell n\left(1 + \hat\partial_\tau^{-1}\frac{\partial^2 S}{\partial\phi^2(\tau)}\right)|\tau >$$

$$= \int d\tau \left[< \tau|\, \hat\partial_\tau^{-1}\frac{\partial^2 S}{\partial\phi^2(\hat\tau)}|\tau > + < \tau|\, \hat\partial_\tau^{-1}\frac{\partial^2 S}{\partial\phi^2(\hat\tau)}\hat\partial_\tau^{-1}\frac{\partial^2 S}{\mu\phi^2(\hat\tau)}|\tau > + \,.....\right] \tag{4.14}$$

With the retarded boundary condition we have

$$< \tau|\, \hat\partial_\tau^{-1}|\tau' > = \theta(\tau-\tau'). \tag{4.15}$$

It is easy to convince ourselves that from second term on are zero. The first term of (4.15) is then

$$\int d\tau d\tau' \; \theta(\tau-\tau') \quad \delta(\tau-\tau') \frac{\delta^2 S}{\delta\phi^2(\tau')} = \underset{\substack{\| \\ 1/2}}{\theta(0)} \int d\tau \; \frac{\partial^2 S}{\partial\phi^2(\tau)} \tag{4.16}$$

Thus, we obtain (4.3).

For the case of advanced boundary condition, the same calculation goes but we must use

$$< \tau | \; \hat{\partial}_\tau^{-1} \; | \tau' > = - \theta(\tau'-\tau) \tag{4.17}$$

accordingly

$$\left|\left|\frac{\delta\eta}{\delta\phi}\right|\right| = \exp\left(- \frac{1}{2} \int d\tau \; \frac{\partial^2 S}{\partial\phi^2(\tau)} \right) \tag{4.18}$$

Hidden Supersymmetry[9]:

Let us go back to the expression (4.2):

$$Z = \int \ldots \int D\phi \; \left|\left|\frac{\delta\eta}{\delta\phi}\right|\right| \; e^{-\int_0^\tau \frac{1}{4} (\dot{\phi} + \frac{\partial S}{\partial\phi})^2 \, d\tau} \tag{4.19}$$

where $\left|\left|\frac{\delta\eta}{\delta\phi}\right|\right|$ is a Jacobian, which depends on the boundary condition imposed on the Langevin equation. One formally replaces this Jacobian by Fermion integration:

$$\left|\left|\frac{\delta\eta}{\delta\phi}\right|\right| = \det \left[(\partial_\tau + \frac{\partial^2 S}{\partial\phi^2(\tau)}) \; \delta(\tau-\tau') \right]$$

$$= \int D\psi D\bar{\psi} \;\; e^{-\int_0^\tau \bar{\psi}(\tau) \, (\partial_\tau + \frac{\partial^2 S}{\partial\phi^2}) \; \psi(\tau) \; d\tau} \tag{4.20}$$

Then the partititon function (4.19) becomes

$$Z = \int \dots \int D\phi D\bar{\psi} \; e^{- \int_0^\tau d\tau \, L} \tag{4.21}$$

where

$$L = \frac{1}{4} \dot{\phi}^2 + \left(\frac{\partial S}{\partial \phi}\right)^2 + \bar{\psi} \left(\partial_\tau + \frac{\partial^2 S}{\partial \phi^2}\right) \psi \tag{4.22}$$

This is the same Lagrangian considered by E. Witten[10] as an example of non-relativistic supersymmetry. The Lagrangian admits a supersymmetry

$$\delta\phi = \bar{\varepsilon}\psi + \varepsilon\bar{\psi}$$

$$\delta\psi = \frac{\varepsilon}{2} \left(\dot{\phi} + \frac{\partial S}{\partial \phi}\right)$$

$$\delta\bar{\psi} = \frac{\bar{\varepsilon}}{2} \left(- \dot{\phi} + \frac{\partial S}{\partial \phi}\right) \tag{4.23}$$

provided the boundary condition of ϕ, ψ and $\bar{\psi}$ are the same. Thus, we impose the periodic boundary condition on the path-integral (4.21) and denote it by Z_{ss}. By using the relation between path-integral and operator expression we obtain

$$Z_{ss} = \mathrm{tr} \, [(-)^F \, e^{- \tau \hat{H}_{ss}}] \tag{4.24}$$

where F is the Fermion number operator and $\hat{H}_{ss}$ is given by

$$\hat{H}_{ss} = \hat{\pi}^2 + \frac{1}{4} \left(\frac{\partial S}{\partial \phi}\right)^2 + \frac{1}{2} [\hat{\psi}^\dagger, \hat{\psi}] \frac{\partial^2 S}{\partial \phi^2} \tag{4.25}$$

Since in the two component vector space of Fermion number 0 and 1, $(-)^F = [\hat{\psi}^\dagger, \hat{\psi}] = \sigma_3$ one obtains

$$Z_{ss} = Z_F - Z_B \tag{4.26}$$

where

$$Z_F = \mathrm{tr}\, e^{-\tau \hat{H}_F}, \; Z_B = \mathrm{tr}\, e^{-\tau \hat{H}_B}$$

$$\hat{H}_F = \hat{Q}^\dagger \hat{Q} \quad , \; \hat{H}_B = \hat{Q}\hat{Q}^\dagger \tag{4.27}$$

Note $\hat{H}_F$ is the same as (4.7).

The energy spectra of $\hat{H}_F$ and $\hat{H}_B$ are semi-positive definite. If zero energy state of $\hat{H}_F$ is normalizable and that of $\hat{H}_B$ is non-normalizable one obtains

$$\lim_{\tau\to\infty} Z_{ss} = \lim_{\tau\to\infty} Z_F \tag{4.28}$$

Thus, only in the case of unbroken supersymmetry the supersymmetric formalism approahces the form of stochastic quantization.

V. STOCHASTIC QUANTIZATION OF GAUGE FIELDS

One of the main reasons condisered by Parisi and Wu[1] to invent the stochastic quantization is the advantage of this method for gauge theories, since the gauge fixing is not necessary for the perturbation calculation. We shall examine this by considering a simple example of a pure Abelian gauge field.

The action is given by

$$S = \frac{1}{4}\int dx\, F^2_{\mu\nu} = \frac{1}{2}\int dx\, A_\mu \,[-\Box\, \delta_{\mu\nu} + \partial_\mu \partial_\nu]\, A_\nu \tag{5.1}$$

The Langevin equation is then obtained by using presctiption of I:

$$\frac{\partial}{\partial\tau}\, \tilde{A}_\mu\,(k,\tau) = -\, k^2(\delta_{\mu\nu} - \frac{k_\mu k_\nu}{k^2}\, \tilde{A}_\nu(k,\tau)) + \tilde{\eta}_\mu(k,\tau) \tag{5.2}$$

where $\tilde{A}_\mu$, $\tilde{\eta}_\mu$.... are Fourier transform of A_μ, η_μ....., respectively.

The solution of Langevin equation (5.2) is given by

$$\tilde{A}_\mu(k,\tau) = \int_0^\tau G_{\mu\nu}(k;\tau-\tau')\,\tilde{\eta}_\nu(k,\tau')\,d\tau' + A^0_\mu(k) \qquad (5.3)$$

where $G_{\mu\nu}$ is a Green's function:

$$[\delta_{\mu\nu}\frac{\partial}{\partial\tau} + (k^2\delta_{\mu\nu} - k_\mu k_\nu)]G_{\nu\sigma}(k;\tau-\tau') = \delta_{\mu\sigma}\,\delta(\tau-\tau') \qquad (5.4)$$

Because of the additional time term it is possible to invert without fixing gauge and we obtain

$$G_{\mu\nu}(k;\tau-\tau') = \{ (\delta_{\mu\nu} - \frac{k_\mu k_\nu}{k^2})e^{-k^2(\tau-\tau')} + \frac{k_\mu k_\nu}{k^2} \}\,\theta(\tau-\tau') \qquad (5.5)$$

In order to obtain a two point correlation function we compute the η- average of $\tilde{A}_\mu(k,\tau)\,\tilde{A}_\nu(k',\tau')$ and obtain

$$\begin{aligned} &< \tilde{A}_\mu(k,\tau)\,\tilde{A}_\mu(k',\tau') >_\eta \\ &= \delta(k+k')\,\{ \frac{1}{k^2}(\delta_{\mu\nu} - \frac{k_\mu k_\nu}{k^2})\,[e^{-k^2|\tau-\tau'|} - e^{-k^2(\tau+\tau')}] + \\ &+ 2\tau_< \frac{k_\mu k_\nu}{k^2} \} \\ &+ \frac{k_\mu k'_\nu}{k^2 k'^2}(k_\mu\,\tilde{A}^0_\mu(k))\,(k_\nu\,\tilde{A}^0_\nu(k)) \end{aligned} \qquad (5.6)$$

where $\tau_<$ stands for the smaller between τ and τ'.

From this expression immediately one can conclude that i) the longitudinal component of two point function diverges at $\tau\to\infty$, ii) the longitudinal component of the initial field configuration remains, iii) the gauge invariant correlation functions such as

$< F_{\mu\nu}(x) F_{\rho\sigma}(y) >$ are finite, and iv) with the choice of $A^0_\mu = 0$ one obtains the Landau gauge results. These are essentially the observation due to Parisi and Wu.[1]

Let us try to understand these features by the path-integral formalism described in IV. The corresponding expression to (4.4) and (4.5) are

$$P[A,t] = e^{-\frac{1}{2} S[A]} < A| e^{-\tau\hat{H}} |A^0 > e^{\frac{1}{2} S[A^0]} \tag{5.7}$$

and

$$< A| e^{-\tau\hat{H}} |A^0 > = \int_{\substack{A_\mu(x,\tau) = A_\mu(x) \\ A_\mu(x,0) = A^0_\mu(x)}} \cdots \int DA_\mu \exp[- \int_0^\tau dx d\tau \{\frac{1}{4} \dot{A}^2_\mu + \frac{1}{4} (\frac{\delta^2 S}{\delta A_\mu}) - \frac{1}{2} \frac{\delta^2 S}{\delta A^2_\mu} \}] \tag{5.8}$$

We note that the last two terms of Fokker Planck Lagrangian in (5.8) are gauge invariant. For Abelian case S is given by (5.1) which contains only the transverse components of field:

$$S = \frac{1}{2} \int d^4k \, k^2 |\tilde{A}^T_\mu(k)|^2 \tag{5.9}$$

where

$$\tilde{A}^T_\mu (k) = (\delta_{\mu\nu} - \frac{k_\mu k_\nu}{k^2}) \tilde{A}_\nu(k)$$

$$A^L_\mu (k) = \frac{k_\mu k_\nu}{k^2} \tilde{A}_\nu(k) \tag{5.10}$$

Therefore the corresponding Fokker Planck Hamiltonian is separated into a sum of longitudinal and transverse part. The longitudinal part does not contain the potential term, namely the longitudinal component is cyclic in the Fokker Planck dynamics:

$$< A|\, e^{-\tau\hat{H}}\, |A^{0}> = < A^{T}|\, e^{-\tau\hat{H}_{T}}\, |A^{OT}>\, < A^{L}|\, e^{-\tau\hat{H}_{L}} |A^{OL}> \tag{5.11}$$

$$\hat{H}_{L} = \int d^{4}k\, |\tilde{\pi}_{L}(k)|^{2} \tag{5.12}$$

In the Langevin language this means there is no drift force.

Since $\hat{H}_L$ is equivalent to a sum of free particle Hamiltonians Feynman kernel is obtained as a product of free particle kernels:

$$< A^{L}|\, e^{-\tau\hat{H}_{L}} |A^{OL}> \; = \text{const}\; e^{-\frac{1}{4\tau}\int d^{4}x\, (A^{L}_{\mu} - A^{OL}_{\mu}(x))^{2}} \tag{5.13}$$

Thus, after a simple calculation one obtains

$$\int DA^{L}\, \tilde{A}^{L}(k)\, \tilde{A}^{L}(-k')\;\; < A^{L}|\, e^{-\tau\hat{H}_{L}}\, |A^{L}_{0}>$$

$$= \; 2\delta_{kk'}\tau + \tilde{A}^{OL}(k)\, \tilde{A}^{OL}(-k') \tag{5.14}$$

The machinery used here is not particularly useful for Abelian case. One can obtain the result much more quickly by using the Langevin equation. For the non-Abelian gauge theories the gauge invariant separation of longitudinal and transverse mode is not possible so that the problem becomes much more complicated. The Fokker Planck formalism used here may be useful in this case because the Fokker Planck Lagrangian resembles the standard non-Abelian gauge theories in $A_0 = 0$ gauge[11] and the collective coordinate technique used for the separation of variables for that problem may be used.

We add here the important obserbations noted by Namiki et al.[12]
i) The result of Landau gauge obtained by Parisi-Wu is due to a specific initial condition ($A^{0}_{\mu} = 0$). Keeping the initial configura-

tion finite and appropriately averaging over it they showed the gauge can be changed. Then the Landau gauge of Parisi and Wu is due to the initial condition. ii) The stochastic quantization of non-Abelian gauge fields yields correctly the effects of Faddeev-Popov ghost fields without introducing them.

Stochastic Gauge Fixing:

It is not necessary to fix gauge in stochastic quantization provided one computes gauge invariant quantities. However, for the purpose of computer simulation it is wise to fix gauge so that damping force acts on the non-gauge invariant modes also. This problem was first considered by Zwanziger[13].

Let F[A] be a gauge invariant quantity of non-Abelian gauge field (such as tr $(F_{\mu\nu}(x)\, F_{\rho\sigma}(x))$, $\mathrm{tr}P\, e^{i\oint A_\mu(x)\, dx_\mu}$ (Wilson loop) etc.:

$$F[A^{\Omega}] = F[A] \tag{5.15}$$

where A^{Ω} is defined by

$$A^{\Omega}_{\mu}(x) = u(\Omega)\, A_{\mu}(x)\, u^{\dagger}(\Omega(x)) + i\, u(\Omega(x))\, \partial_{\mu} u^{\dagger}(\Omega(x)) \tag{5.16}$$

The average of F[A] at τ is given by

$$\begin{aligned} < F[A] >_{\tau} &\equiv \int \cdots \int DA\, F[A]\, P[A,\tau] \\ &= \int \cdots \int DA\, F[A]\, < A|\, e^{-\tau \tilde{H}_{FP}} |A^{0}> \end{aligned} \tag{5.17}$$

where

$$\tilde{H}_{FP} = - \int dx\, \frac{\delta}{\delta A^{a}_{\mu}(x)} \left(\frac{\delta}{\delta A^{a}_{\mu}(x)} + \frac{\delta S[A]}{\delta\ _{\mu}(x)} \right) \tag{5.18}$$

The measure DA in (5.17) is gauge invariant. Thus, one obtains

$$< F[A] >_\tau = \int..\int DA\ F[A]\ < A^\Omega |\ e^{-\tau \tilde{H}_{FP}}\ |A^0 >$$

$$= \int..\int DA\ F[A] < A|\ e^{-i\int G_a(x)\Omega^a(x,\tau)dx}\ e^{-\tau \tilde{H}_{FP}}\ |A^0 > \qquad (5.19)$$

where $G_a(x)$ is a generator of the gauge transformation.
Since the operators inside the bracket can be obtained by an appropriate τ dependent gauge transformation from $e^{-\tau \tilde{H}_{FP}}$, the corresponding Langevin equation is also obtained by this τ dependent gauge transformation[14].

$$\frac{\partial}{\partial \tau} A_\mu(x,\tau) = - \frac{\delta S}{\delta A_\mu} + D_\mu v + \eta_\mu \qquad (5.20)$$

where

$$v = - i \left(\frac{\partial}{\partial \tau} u(\Omega(x,\tau))\right) u^{-1} \qquad (5.21)$$

Since Ω is arbitrary v is arbitrary. The second term (5.20) is the desired damping force along the gauge orbits.

VI. FURTHER REMARKS

The stochastic quantization reviewed in the previous sections is another quantization method which is designed to be equivalent to the Euclidean path-integral quantization. It is useful for the numerical simulations, although it does not give any great advantage compared to other methods[15]. Theoretically, however, the stochastic quantization seems to provide some advantages to some problems such as gauge theories discussed in the previous section. Further, it was shown that the quenched Eguchi-Kawai model at large N can be derived more elegantly by this method[16]. Nonetheless, I must say that no essentially new results came out of this method yet.

The other topics studied in stochastic quantization are the stochastic regularization and renormalization. Keeping τ finite makes the theory regularize to some extent but this is not sufficient. Further regularization is necessary. The stochastic regularization was proposed by Breit et al[4]. In the η average one does not take $\Lambda \to \infty$ and keep finite. This regularizes the expression finite. Since this regulates only the fictitious time direction it does not affect invariance of the theory such as gauge and Lorentz invariance. A systematic study of this regularization with finite τ and the renormalization was done by Alfaro[17]. He concludes that the stochastic regularization works for renormalizable theories but for unrenormalizable theories not only the stochastic regularization does not work but also the stochastic quantization itself is inconsistent.

An interesting question still unsolved is the question raised by Parisi[18]: "Can stochastic quantization be generalized to complex S?" He suggested the use of Langevin equation for complex S by considering a complex valued probability density. This problem had already been considered by Klauder[19] to some extent in the study of coherent-state Langevin equation. Let us elaborate the point.

Let us consider the Euclidean path-integral:

$$Z = \int .. \int Dq \;\; e^{-S[q]} \tag{6.1}$$

where

$$S = \int d\tau \, [\tfrac{1}{2} \dot{q}^2 + V(q)] \tag{6.2}$$

One can develop the stochastic quantization. No problem. However, if one expresses (6.1) in terms of the Euclidean phase space path-integral

$$Z = \int .. \int DpDq \, e^{-S[p,q]} \tag{6.3}$$

$$S[p,q] = \int d\tau \, [-ip\dot{q} + \tfrac{1}{2} p^2 + V(q)] \tag{6.4}$$

One finds a complex S, accordingly complex probability problem. However, this is merely an artifact due to a Fourier transform of the Gaussian integral. Therefore, this kind of difficulty associated with the complex probabilities should be resolved. If one goes back to the discussion of Fermi fields in III, one realizes a similar complex probability problem appeared then already in a different form. However, there exists no general study of this problem yet to my knowledge.

REFERENCES

1. G. Parisi and Wu Yong-Shi, Scientia Sinica, 24, 483 (1981).
2. B. Sakita, 7th Johns Hopkins Workshop, ed. G. Domokos and S. Kovesi-Domokos (World Scientific, 1983).
3. M. Namiki and Y. Yamanaka, Prog. Theor. Phys. 69, 1764 (1983).
4. J.D. Breit, S. Gupta and A. Zaks, Nucl. Phys. B233, 61 (1984).
5. G. Parisi (unpublished)
   I.T. Drummond, S. Duane, R.R. Gogan, Nucl. Phys. B220, 119 (1983).
   A. Guha and S.-C. Lee, Phys. Lett. 134B, 216 (1984).
   J. Alfaro and B. Sakita, Proc. Topical Symp. High Energy Phys., ed. T. Eguchi and Y. Yamaguchi (World Scientific, 1983).
   M.B. Halpern, UCB-PTH - 83/1.
6. There were errors in reference 2. The editors disregarded the corrections sent a few days later.] I corrected them here. Similar Langevin euations were written and justified by perturbative calculations:
   J.D. Breit, S. Gupta and A. Zaks, Nucl. Phys. B233, 61 (1984).
   P.M. Damgaard, K. Tsokos, Nucl. Phys. B235, 75 (1984).
7. T. Fukai, H. Nakazato, I Ohba, K. Okano, Y. Yamahaka, Prog. Theor. Phys. 69, 1600 (1983).
8. E. Gozzi, Phys. Rev. D28, 1922 (1983).
   See also B. Sakita, Proc. Leipzig Conference (1984).
9. G. Parisi and N. Sourlas, Phys. Rev. Lett. 43, 744 (1979); Nucl. Phys. B206, 321 (1982).
   E. Gozzi, Phys. Rev. D28, 1922 (1983).
   E.S. Egorian, S. Kalitzin, Phys. Lett. 129B, 320 (1983).

R. Kirshner, Phys. Lett. B139, 180 (1984).

10. E. Witten, Nucl. Phys. 188B, 513 (1981).
11. J.L. Gervais and B. Sakita, Phys. Rev. D18, 453 (1978).
12. M. Namiki, I. Ohba, K. Okano, Y. Yamanaka, Prog. Theor. Phys. 69, 1580 (1983).
13. D. Zwanziger, Nucl. Phys. B192, 259 (1981).
14. M. Horibe, A. Hosoya, J. Sakamoto, Prog. Theor. Phys. 70, 1636 (1983).
15. A. Guha, Talk given in this Workshop.
16. J. Alfaro and B. Sakita, Phys. Lett. 121B, 339 (1983).
    J. Alfaro, Phys. Rev. D28, 1001 (1983).
    J. Greensite and M.B. Halpern, Nucl. Phys. B211, 339 (1983).
17. J. Alfaro, LPTENS 84/4 - Trieste preprint (July 1984).
18. G. Parisi, Phys. Lett. 131B, 393 (1983).
19. J.R. Klauder, Acta Physica Austriaca, Suppl. XXV251 (1983); Phys. Rev. A29, 2036 (1984).

Notes

In the preparation of the lectures the following references were used.

II.

L.D. Faddeev, "Introduction to Functional Methods" Les Houches 1975, Session 28, North-Holland, 1976

III.

The notion of integration over the Grassmann number was introduced by Berezin.

F.A. Berezin, Method Second Quantization, Academic Press, New York, 1966.

But Faddeev's lecture note contains the necessary information.

IV.

C. Itzykson and J.B. Zuber, Quantum Field Theory, McGraw-Hill, New York 1980.

V.

R.P. Feynman, Statistical Mechanics, Benjamin, Mass. 1972.
F. Takano, Many Body Problems, Baifukan, Tokyo, 1975 (in Japanese)
L.P. Gorkov, JETP, 7, 505 (1958).
H.B. Nielsen and P. Olesen, Nucl. Phys. B61, 45 (1973).

VI.

J.-L. Gervais and A. Jevicki, Nucl. Phys. B110, 93 (1976).
M. Sato, Prog. Theor. Phys. 58, 1262 (1977).

VII.

The method presented in this chapter is based on

A. Jevicki and B. Sakita, Nucl. Phys. B165, 511 (1980).

The collective excitations of SU(N) symmetric matrix model discussed in the last section is due to

M. Mondello and E. Onofri, Phys. Letters 98B, 277 (1981)

The large N planar limit of SU(N) was investigated earlier by

E. Brezin, C. Itzykson, G. Parisi and J.B. Zuber, Comm. Math. Phys. 59, 35 (1978).

VIII.

R.P. Feynman, Statistical Mechanics, Benjamin, Mass (1972).

The discussion presented here is based on

B. Sakita,"Intermediate Coupling Theory", Proc. Brown Workshop on Nonperturbative Studies in Quantum Chromodynamics, 1981.

Lee-Low-Pines theory:

T.D. Lee, F.E. Low, D. Pines, Phys. Rev. 90, 297 (1953).

Intermediate Coupling theory:

S. Tomonaga, Prog. Theor. Phys. 2, 6 (1947).

IX.

H. de Vega, J.-L Gervais and B. Sakita, Nucl. Phys. B139 20 (1978); Nucl. Phys. B142, 125 (1978).

X.

J.-L. Gervais, A. Jevicki, B. Sakita, Phys. Rev. D12, 1038 (1975).
J.-L. Gervais, A. Jevicki, Nucl. Phys. B110, 13 (1976).
R. Dashen, B. Hasslaher, A. Neveu, Phys. Rev. D10, 4114 (1974);

D10, 4130 (1974); D11, 3424 (1975).

AI.

J.-L. Gervais and B. Sakita, Phys. Rev. D18, 453 (1978).
N.H. Christ, T.D. Lee, Phys. Rev. D22, 939 (1980).

AII.

C.J. Hamer, J.B. Kogut, L. Susskind, Phys. Rev. D19, 3019 (1979).
J.B. Kogut, L. Susskind, Phys. Rev. D11, 395 (1975).

Appendix AI and AII are a part of the lecture notes published in Japanese in Soken 62, 214-259 (1981).

Index